International Congress on Energy 2018

Core Programming Area at the 2018 AIChE Annual Meeting

Pittsburgh, Pennsylvania, USA
28 October - 2 November 2018

ISBN: 978-1-5108-7608-8

Printed from e-media with permission by:

Curran Associates, Inc.
57 Morehouse Lane
Red Hook, NY 12571

Some format issues inherent in the e-media version may also appear in this print version.

Printed by Curran Associates, Inc. (2019)

AIChE
120 Wall Street, FL 23
New York, NY 10005-4020

Phone: (800) 242-4363
Fax: (203) 775-5177

www.aiche.org

Additional copies of this publication are available from:

Curran Associates, Inc.
57 Morehouse Lane
Red Hook, NY 12571 USA
Phone: 845-758-0400
Fax: 845-758-2633
Email: curran@proceedings.com
Web: www.proceedings.com

TABLE OF CONTENTS

Some abstracts in this book refer to images that were not made available by the publisher

526363 Experimental and Theoretical Studies of the Electrochemical Reduction of CO_2 on Cu

Abstract:

The electrochemical and photoelectrochemical reduction of CO_2 to chemicals and fuels has been the subject of considerable interest in recent years. We have been pursuing the subject through a series of experimental and theoretical studies with the objective of identifying the factors controlling the distribution of products formed, with particular interest on establishing how to enhance the formation of C_{2+} hydrocarbons and oxygenates. This work has revealed that the activity of Cu for forming such products depends on the surface composition and morphology of Cu, the composition and concentration of electrolyte cation and anions, and the magnitude of the applied potential. We have also found that mass transfer of CO_2 to the catalyst surface and OH^- anions from the catalyst surface can also modify the distribution of products observed. This talk will talk will illustrate how information obtained from experiments and theory can be used to elucidate the fundamental processes governing product distribution.

538675 Highly Efficient CO Electroreduction Catalyst Based on Polycrystalline Cu Particles

Abstract:

Electrochemical conversion CO_2 to value-added chemicals and fuels using renewable energy is a promising technique for energy storage and to potentially diminish problems associated with CO_2 emissions. Many catalysts are able to reduce CO_2 to CO with high selectivity and activity. However, to achieve more energy-dense products requires further reduction of CO. Cu is the only currently known material exhibiting appreciable activities towards CO electroreduction[1,2]. Recent studies show grain boundary terminated Cu surface is able to effectively improve the selectivity and required energy bias for CO electroreduction[2,3]. Despite significant progress, the optimal activity achieved is only less than 1 mA/cm^2 which is attributed to the poor transportation of CO caused by its low solubility in the electrolyte (~1 mM). Herein we report our latest progress in CO electroreduction using polycrystalline Cu particle catalysts to achieve high efficiency. Our efforts in providing mechanistic insights including reaction intermediates and pathways will also be discussed.

[1] Hori Y, Takahashi R, Yoshinami Y, et al. The Journal of Physical Chemistry B, 1997, 101(36): 7075-7081.

[2] Li C W, Ciston J, Kanan M W. Nature, 2014, 508(7497): 504.

[3] Verdaguer-Casadevall A, Li C W, Johansson T P, et al. Journal of the American Chemical Society, 2015, 137(31): 9808-9811.

538294 Operando Spectroscopic Investigations of Oxide Derived Metal Catalysts for CO_2 and CO Reduction

Abstract:

Electrochemical reduction of CO_2 driven by renewable electricity is a promising approach to convert CO_2 into valuable chemicals and fuels. Copper is the only metal catalyst for this reaction with significant selectivity towards hydrocarbons and oxygenates. CO is a known intermediate in the reduction of CO_2, and thus its conversion is also of notable interest. Oxide derived copper (OD-Cu) has been shown to favor the more desirable C2 and C3 products at lower overpotentials than bulk copper. Despite numerous studies and proposed mechanisms, e.g., the presence of grain boundaries, surface pH and surface roughness, the exact nature of the active phase is still a topic of discussion. In this work, we use operando attenuated total reflectance surface enhanced infrared reflection absorption spectroscopy (ATR-SEIRAS) to probe different types of sites available on Cu and OD-Cu surfaces using CO as a probe molecule. We show that multiple types of CO-binding sites are available on both Cu and OD-Cu surfaces. Further, the types of binding sites present are also strongly potential dependent. We identify strong correlations between the presence of several types of surface sites with the ability of the catalyst to produce C2 and C3 products at -0.4 V vs. RHE.

531291 Molecular-Level Insights into Electrocatalytic Carbon Dioxide Reduction at Cobalt Macrocycles

Abstract:

Electrolysis of carbon dioxide in aqueous media to generate carbon monoxide may provide a modular route to making a key chemical feedstock under ambient conditions. As a component of syngas, carbon monoxide can be converted into higher value hydrocarbons, alcohols, acids, and esters. Cobalt macrocycles, including cobalt phthalocyanines and cobalt porphyrins, are known to be selective catalysts for converting carbon dioxide into carbon monoxide. In our work, we have been able to develop immobilized versions of these catalysts with high turnover frequencies for reducing carbon dioxide. These catalysts have helped to elucidate a rich picture of the nature of the rate determining step, how it can be tuned between an electron transfer and proton-coupled electron transfer, and the interplay of kinetics and transport in the observed reactivity of cobalt macrocycles. Furthermore, we have been able to deconvolute both inductive and electrostatic effects on the activity of these catalysts, providing clear design principles for transition metal macrocycles which are immobilized at electrode-electrolyte interfaces.

527408 Mass Transfer Effects in CO_2 Reduction Electrocatalysis

Abstract:

Significant interests have risen in the development of catalytic nanomaterials for the electroreduction of CO_2. While extensive studies have been reported on tuning of surface structures to improve the chemical kinetics, less attention has been paid to the mass transfer effects in the CO_2 reduction reaction on nanoscale electrocatalysts. We report here an systematic investigation of CO_2 electroreduction on highly dense Cu nanowires, with the focus placed on practical high-flux conditions. Mass transfer effects are found to play an important role in this case, giving rise to diffusion-limited CO_2 reduction activity and selectivity. By correlating the observed transport phenomena to the calculated CO_2 consumption rate and simulated surface concentration of chemical species (including local pH) on the nanowires, an upper limit is revealed for the CO_2 conversion rate on the nanostructured electrodes, which also causes the drop in Faradaic efficiency of CO_2 reduction at large current densities. Our work emphasizes the necessity of considering mass transfer effects in the design of advanced electrocatalysts for CO_2 reduction as well as for understanding their structure-performance relationships.

536098 Single Atom Catalysts for Electrochemical Reduction of CO_2

Abstract:

Electrochemical reduction of CO_2 offers a sustainable route to mitigate the problem of rising in CO_2 concentration in the atmosphere. Among the various electrocatalysts studied for the conversion of CO_2 to hydrocarbons, supported metal catalysts have played an important role in enhancing the rate of chemical reactions compared to bulk unsupported metals. Broadening this distribution of metal to make supported isolated atoms is a novel approach to heterogeneous catalysis. The reduction in the size of the catalyst not only enlarges its specific area, but also increases the number of under coordinated sites, and the HOMO-LUMO gap. This can improve the activity and selectivity of the catalyst. In this work, Î·^{6}-functionalized graphene which includes transition metal single-atom (SA) layer is used as the cathode for electrochemical reduction of CO_2. The Î·^{6} chemistry retains grapheneâ€™s 2D structure and high electronic properties (high carrier mobility and high electric conductivity), which contributes to a high catalytic activity with a large surface area. Moreover, the transition metal is grafted on graphene by coordination bonds, so that the adsorption energy of CO_2 reduction intermediates is increased and further the energy barrier for hydrogenation is reduced. Such single metal atom supported electrocatalysts offer a higher activity to transition metals as compared to their bulk counterparts. The use of SA catalysts also means that lesser quantities of metals being used per unit weight of the electrode resulting in cheaper catalysts with superior performance. Due to only recent advances in the synthesis of stable SA catalysts, the application of single-atom catalysis for electrochemical reduction of CO_2 seems promising. To study the behavior of single-atom catalysts for CO_2 reduction, we present a design of a compact electrochemical cell with carbon supported single-atom transition metal atoms on graphene as the Rotating Disk Electrode (RDE). An RDE is chosen as it minimizes the cathode polarization and enables the measurement of intrinsic kinetic activity. A product distribution of CO_2 reduction products at near-neutral pH, their partial current density, and Faradaic efficiencies at different potentials will also be reported in this work.

524710 Photoelectrochemical CO_2 Reduction at Plasmonic Nanostructured Silver Electrodes

Abstract:

Electrochemical CO_2 reduction could help mitigate carbon emissions while producing valuable fuels or chemical feedstocks. However, inefficiency and low selectivity toward a hydrocarbon product have limited the commercial viability of this process. Plasmonic hot-carriers and the strong local electric fields produced by plasmon excitation may open new pathways for CO_2 reduction, resulting in increased product selectivity and lower overpotential. Our research aims to understand the fundamentals underlying plasmon-generated excited charge transfer from voltage-biased cathodes to CO_2 reduction intermediates to selectively and efficiently form desirable hydrocarbon products.

We have designed a custom gas flow cell to probe the mechanism of plasmon-enhanced CO_2 reduction on nanostructured noble metal electrodes fabricated by nanosphere lithography. This temperature-controlled photoelectrochemical cell is specifically designed to allow quantitative analysis of gaseous and liquid products formed at an illuminated working electrode. The geometry of the cell enables front-illumination of the photoelectrode and maximizes the electrode surface area to electrolyte volume ratio to increase liquid product concentration and hence enhance *ex situ* spectroscopic sensitivity towards them. Gas is bubbled through the electrolyte in the working electrode chamber during operation to maintain a saturated reactant concentration and to continuously mix the electrolyte. Gaseous products are detected by an in-line gas chromatograph, and liquid products are analyzed *ex situ* by NMR. Product distribution and photocurrent are measured while varying electrochemical potential, illumination intensity, light wavelength, temperature, and the plasmonic catalyst. We demonstrate that plasmonic photocurrent on voltage-biased silver (Ag) nanopyramid cathodes is selective for CO_2 reduction over the competing hydrogen evolution reaction. Our results suggest that further plasmonic enhancements in selectivity and activity towards specific CO_2 reduction reactions are possible by tuning the electrode structure and composition.

This material is based upon work performed by the Joint Center for Artificial Photosynthesis, a DOE Energy Innovation Hub, supported through the Office of Science of the U.S. Department of Energy under Award Number DE-SC0004993.

The work is supported by the National Science Foundation Graduate Research Fellowship under Grant No. 1106400.

Work at the Molecular Foundry was supported by the Office of Science, Office of Basic Energy Sciences, of the U.S. Department of Energy under Contract No. DE-AC02-05CH11231.

540773 2018 Outlook for Energy: A View to 2040

Abstract:

The Outlook for Energy is ExxonMobil's global view of energy demand and supply through 2040. It also highlights the dual challenge of providing affordable energy to support prosperity while reducing environmental impacts – including the risks of climate change. We use the Outlook to help inform our long-term business strategies and investment plans.
The Outlook anticipates significant changes through 2040 across the world to boost living standards, reshape the use of energy, broaden access to abundant energy supplies, and accelerate decarbonization of the world's energy system to address the risks of climate change.

- Global energy needs will rise about 25 percent over the period to 2040
- Electricity demand will nearly double in non-OECD countries
- Worldwide electricity from solar and wind will increase about 400 percent
- Natural gas will expand its role, led by growth for electricity generation
- Growth in oil demand will be driven by commercial transportation and chemical needs
- Efficiency gains and growing use of less-carbon intensive energy sources will contribute to a nearly 45 percent decline in the carbon intensity of global GDP

Energy is fundamental to modern life, and as the world's population approaches 9 billion people in 2040, we are challenged to improve living standards everywhere. Understanding the factors that drive the world's energy needs – and likely choices to meet those needs – is the mission of the Outlook. By sharing the Outlook with the public, we hope to broaden that understanding. Energy matters to everyone, and we all play a role in shaping its future.

541543 Energy Decarbonisation Scenarios

Abstract:

With the energy sector being associated with 2/3 of the annual greenhouse gas emissions, decarbonizing the energy is a key step to reduce their level. The National Determined Contributions (NDC's) announced so far by the participating countries do not meet the objectives of the Paris-climate agreement, as far as reaching a peak in emissions in the near future or decreasing to a level that is compatible with the stated long-term targets.

Using different scenarios for energy demand and supply, options for energy sector decarbonisation are evaluated. Meeting the 2-degree target requires a major transformation in the power sector, efficiency improvements – specially in the buildings and in the industry, and a revolution in transport towards fuel and modal shifts. With a significant increase in renewable energy, system integration and optimization will also be key in reducing the cost of the energy transition. Meeting more ambitious climate goals ("well below 2 degrees") will push further the technological innovation and implementation requirements throughout the entire energy value chain. A scenario aiming at zero net-emissions by 2060 is presented, along with a comparison to the 2-degree case.

542625 Fundamental Research Needs to Advance Energy Technologies

Abstract:

The U. S. Department of Energy (DOE), Office of Science, Office of Basic Energy Sciences (BES) supports fundamental research in chemical and materials sciences to provide the foundations for new energy technologies and to support DOE missions in energy, environment, and national security. This presentation will discuss opportunities for fundamental research to impact DOE's energy mission "to catalyze the timely, material, and efficient transformation of the nation's energy system and secure U.S. leadership in energy technologies" with a focus on the way we generate, store and use energy nationally. I will provide an overview of BES strategic planning over the past decade that identified priority research directions for advancing energy applications, highlight key scientific advances in these areas, and discuss some future opportunities for modern science, particularly chemical sciences, to accelerate the transformation of the U. S. energy portfolio.

531924 Solid-State Depolymerization and Isolation of Lignin from Lignocellulosic Biomass

Abstract:

This study demonstrated that lignin could be efficiently depolymerized in the solid state, with minimal condensation, and separated from dissolved and hydrolyzed cellulose and hemicelluloses, in an inorganic ionic liquid (acidic lithium bromide trihydrate (ALBTH) system) under mild conditions (with 40 mM HCl at 110 °C). The resultant ALBTH lignin contained abundant uncondensed moieties (i.e., Hibbert's ketones and benzodioxanes), from partial depolymerization, and could readily be isolated from biomass. Reactions using lignin model compounds confirmed the formation of the uncondensed moieties and revealed the synergy between LiBr and acid in inducing the crucial intermediate α-benzyl carbocations, which then led to cleavage of the β–*O*–4-aryl ether bonds to produce Hibbert's ketones, demethylation to produce benzodioxanes, and condensation reactions. Unlike lignin model compounds, the condensation of the real lignin in biomass under ALBTH conditions was greatly diminished, possibly due to lignin's remaining in the solid state, limiting its mobility and accessibility of the a-benzyl carbocation to the electron-rich aromatic sites for condensation. This study provided a new approach to effectively isolate depolymerized lignin from lignocellulose in a less condensed form for boosting its downstream valorization.

539017 Catalytic Depolymerization and Liquefaction of Lignin in Ionic Liquid By SO_4^{2-}/ZrO_2 in a Flow through System

Abstract:

Lignin has been considered as a promising sustainable feedstock for high value-added aromatic chemical production. Here, we proposed a novel and e cient process for alkali lignin depolymerization and liquefaction in a flow through system with pretreatment unit, reaction unit and separation unit. In pretreatment unit, lignin was dissolved in N-allyl pyridinium chloride ([Apy]Cl) and SO_4^{2-}/ZrO_2 was used as the catalyst in reaction unit. The effects of different reaction temperatures (150, 170, 190, 210) on the yields of carbon and monomer (on the lignin intake) were investigated, the yields of carbon and monomers appear to be minimized (17.888% and 3.278% respectively) at 150 and increased to 39.179% and 20.254% at 210 respectively. The structural evolution of the alkali lignin was intensively investigated via comparative characterization of raw and residues using FT-IR. The structures of residues were changed with temperature increased, and coke had probably been formed.The thermal stability of [Apy]Cl and regenerated [Apy]Cl were studied. According to the results, during the pretreatment process the β–O–4 linkages were partially cleaved, and in reaction process, SO_4^{2-}/ZrO_2 mainly cleaved the β–O–4, α–O–4 α–O–γ and C–C linkages in lignin and produced phenolic products and their derivatives. Proposed [Apy]Cl pretreatment mechanism of lignin and depolymerization liquefaction mechanism of lignin in [Apy]Cl catalyzed by SO_4^{2-}/ZrO_2 were speculated.

536971 The Impact of Acid Site Concentration and Pore Diameter on the Cracking of Lignin Derived Monomers in Zeolites

Abstract:

In the forest bioproducts industry, tens of millions of metric tons of pulp are produced every year. An important byproduct of the pulping process is lignin, the irregular aromatic biopolymer that acts as a natural glue to hold plants together. Currently, most of the industry burns this complex biopolymer for energy rather than attempt to salvage the valuable aromatic carbon inside. One of the major hurdles to commercializing lignin based organics is that during depolymerization, lignin breaks down into a complex mixture of products. To make lignin valorization an economically feasible process, we must find a way to convert this complex stream into easily separable products.

Using lessons learned from petroleum refining, we show how a mixture of lignin monomers can be converted to an easily separable stream of light hydrocarbons and oxygenated aromatic rings by removing alkyl site chain through cracking reactions. Specifically, this talk focuses on the cracking of propylene from a phenol ring over various types of zeolites. We explore how different aspects of the catalyst, such as the concentration of Brønsted and Lewis acid sites, micropore diameter, and hierarchical structure impact conversion, selectivity, and deactivation, as well as the required regeneration procedure for restoring the original activity. We suggest that strongly bound phenolate species form on the Lewis acid sites, which hinder diffusion of reactants and products into and out of the particles. By selectively removing these sites, we show a change in reactivity and selectivity of the cracking products. Using operando IR spectroscopy, we study how these compounds interact with our catalyst surface and how the various surface species evolve over time, helping us design better and more reliable zeolites.

539104 High Energy Density Fuels Produced from Lignin-Derived Intermediates and Refinery Waste Gas Streams

Abstract:

This paper presents an economic pathway for high energy density fuels production from lignin-derived intermediates and refinery waste gas streams. The lignin-derived intermediates, including toluene and xylenes, are obtained from one-pot lignin conversion catalyzed by metal(s)-zeolite Y catalysts with the presence of hydrogen in water at 250Â°C [1-2]. Alkylation of the intermediates with refinery waste gas streams, which contain up to 10% olefins, and subsequent distillation, produces C8-C12 hydrocarbons with high octane number (>120) and energy density using the existing refinery infrastructure. The octane numbers and energy density of the components are summarized. Brief economic analysis is provided.

[1] H. Wang, H. Ruan, M. Feng, et al, One-Pot Process for Hydrodeoxygenation of Lignin to Alkanes Using Ruâ€Based Bimetallic and Bifunctional Catalysts Supported on Zeolite Y, *ChemSusChem* 2017, 10(8), 1846-1856.

[2] H. Wang, M. Feng, B. Yang, Catalytic hydrodeoxygenation of anisole: an insight into the role of metals in transalkylation reactions in bio-oil upgrading, *Green Chemistry* 2017, 19, 1668-1673.

535560 Towards Valorization of Biorefinery Waste to Polyhydroxyalkanoate: Structural Characterization and Mechanisms

Abstract:

The large-scale production of cellulosic ethanol from lignocellulosic biomass are generating lignins which have high potential to be transformed to sustainable high-value products. Sustainable production of polyhydroxyalkanoate (PHA) using lignin as carbon source provides a promising valorization pathway of lignin. In this study, four lignin streams generated from different fractionation strategies (integrating acid/alkali pretreatment, enzyme hydrolysis, and laccase pretreatment) from corn stover were used in fed-batch fermentation to produce PHA by *P. putida* KT2440. The lignin streams before and after subjecting to the fermentation process as well as the native lignin from corn stover were thoroughly characterized by gel permeation chromatography (GPC), ^{31}P nuclear magnetic resonance (NMR), two-dimensional (2D) heteronuclear single-quantum coherence (HSQC) NMR spectroscopy, and ^{19}F NMR. The GPC data showed that the fractionation processes significantly lower the molecular weight of lignin steams; the polydispersity (PDI) of the lignin streams decreased after the fermentation step with a highest reduction percentage of 50.8%, which may suggest that the lower molecular weight components were more favorable for *P. putida* KT2440 and the molecular weight distribution of lignin streams became uniform. ^{31}P NMR analysis examined the hydroxyl group changes after the fractionation and fermentation processes. Semi-quantitative HSQC data proved that the *p*-hydroxyphenyl lignin may promote the lignin conversion to PHA; more than 78% *p*-coumarate and 15% ferulate structures in lignin steams from three fractionation strategies were consumed, which suggest that consumption of hydroxycinnamates contributed to the production of PHA. ^{19}F NMR analysis achieved the quantification of carbonyl groups in lignin stream; the fermentation generally consumed more than 50.4% aldehyde, ketone, and quinone structures in lignin steam.

538284 Characterization of Deep Eutectic Solvent Extracted Lignin Streams from Endocarp Biomass

Abstract:

Lignin is a promising direct source and contains upgradable building blocks of valuable aromatic chemicals and materials. Endocarp biomass represents a non-edible crop residue in an existing agricultural setting which cannot be used as animal feed nor soil amendment. With considerably higher lignin content and bulk energy density, endocarps have significant advantages to be converted into both biofuel and bioproducts as compared with other biomass resources. As a new pretreatment method, deep eutectic solvent (DES) pretreatment is shown to be highly effective in fractionating lignin from biomass with high yield and purity, in addition to merits of both environmental friendliness and low cost compared with several other pretreatment methods. However, the impact of DES pretreatment on the physical and chemical properties of extracted lignin is so far not clear. Therefore, this study aims to fractionate and characterize lignin streams from endocarps of peach and black walnut using DES in comparison with two conventional pretreatment methods, i.e., dilute acid and alkaline pretreatments. SEM and confocal microscopic analysis revealed distinct structural characteristics of peach and walnut endocarp biomass as compared to switchgrass and woody biomass. Chromatographic and nuclear magnetic resonance (NMR) spectroscopic techniques were employed to understand the lignin composition, structure and depolymerization during DES pretreatment. The resulting lignins were further characterized for purity, molecular weight distribution, and thermogravimetric properties. This study provides a foundation for developing novel conversion techniques for upgrading DES extracted endocarp lignin for value-added chemicals and materials.

538719 Reactivity-Based Fractionation of Lignins Via Reversible Conjunction to Polymeric Amines

Abstract:

Fractionation is a necessary step in the process of upgrading lignins to synthetic materials. Existing methods for separating lignin fractions based on differences in solubility or molecular size generate polymer mixtures with heterogeneous physiochemical properties, which are not suitable for use in high-value plastics or composites. We demonstrate the utility of catalytic oxidation as an approach to partially depolymerizing lignins from diverse sources including raw biomass, enzymatic lignin, and fungal decay lignin. The resulting lignin fragments contain reactive carbonyl groups (i.e. aldehydes and ketones), which conjugate reversibly with primary amines. Using an amine-functionalized polymer as the conjugating reagent, we are able to isolate lignin fractions that are enriched in carbonyl structures. Recovery and reuse of the conjugating reagent are achieved via acid-catalyzed hydrolysis of amine-carbonyl adducts.

538956 Understanding and Modeling Effects of Nitrogen Source on Biosynthesis of Polyhydroxyalkanoates from Benzoate By *Pseudomonas Putida* KT2440

Abstract:

Benzoate has been widely used as a model compound for the study of bacteria catabolism of aromatic compounds. Nitrogen source is critical to the synthesis of intracellular polyhydroxyalkanoates (PHAs) in various bacteria. However, exact effects of nitrogen source in the PHAs biosynthesis process from benzoate in *Psuedomonas putida* is unclear. In this study, the PHAs biosynthesis process from benzoate by *Pseudomonas putida* KT2440 was investigated under nitrogen surplus and limit conditions. In nitrogen limit condition, PHAs accumulated up to 37.3% of its cell dry weight (CDW) and eleven types of monomers were identified through GC/MS analysis. On the contrary, PHAs only accounted for 0.25% of CDW and three main PHAs monomers were found under nitrogen surplus condition. NMR and GPC results confirmed that PHAs molecular weight started to increase when the nitrogen source exhausted while the PHAs concentration maintained at the similar level under surplus nitrogen condition. Kinetic models were developed to adequately predict the biomass growth, PHA accumulation, and benzoate consumption. Results indicated that the cell growth and PHA accumulation fit the Monod-type model and the PHA synthesis process was dependent to nitrogen exhausting condition.

538379 Electrochemical CO_2 Conversion to Valuable Chemicals

Abstract:

Converting greenhouse gas carbon dioxide (CO_2) to value-added chemicals is an appealing approach to tackle CO_2 emission challenges. The chemical transformation of CO_2 requires suitable catalysts that can lower the activation energy barrier to minimize the energy penalty associated with the CO_2 reduction reaction. Cu is the only monometallic catalyst that can produce an appreciable amount of C2+ products from CO2, while the C2+ selectivity of Cu must be further improved in order to be considered for commercialization. Here, we will present our work in developing an integrated two-stage electrolyzer stack system for the conversion of CO_2 to alcohols. The first stage utilizes a silver catalyst in a flow cell capable of reducing CO_2 to CO with faradaic efficiencies of >95% and partial current densities of >200 mA/cm^2. For the second stage, a CO reduction flow cell using alkaline electrolyte can reduce CO to C2+ products with high selectivity and high current density. This work demonstrates the feasible production of multi-carbon alcohols via CO reduction at appreciable rates, and motivates further development of a two-stage electroreduction process of CO_2 to alcohols.

545759 Development of Devices and Selective Catalysts for the Solar-Driven Reduction of CO_2 to fuels

Abstract:

The electrochemical and photoelectrochemical reduction of CO_2 presents an attractive route towards fuel formation from renewable energy. However, the dearth of selective, active and inexpensive catalyst materials remains a key challenge on the way to renewable production of hydrocarbon fuels. Here, we address these challenges through the design of novel devices and selective earth-abundant catalysts for the electrochemical[1–3] and solar-driven[1,4–6] reduction of CO_2. In addition, we provide insight into the origin of selectivity on electrocatalysts based on copper, paving the way towards more rational design of electrochemical fuel forming devices.

We pioneer the application of Atomic Layer Deposition (ALD) to direct the catalytic selectivity of copper catalysts. Copper-based catalysts derived from the reduction of CuO are known to reduce CO_2 but simultaneously exhibit poor selectivity. ALD coating of CuO with thin layers of SnO_2 allows for formation of CO with high selectivity even at low overpotentials, comparing favorably to conventional gold catalysts, known to selectively convert CO_2 to CO. We use kinetic studies, high resolution transmission electron microscopy and gas chemisorption analysis to explain the observed selectivity changes on the basis of decreased binding strengths for both H and CO intermediates on the Sn-containing catalyst surface.

We subsequently combine our new catalyst with an earth abundant oxide anode to effect overall splitting of CO_2 to CO and O_2. Driven by a 3-junction solar cell, this device achieved long-term solar splitting of CO_2 into CO and O_2 at an efficiency of 13.4 %, constituting the current world record.[1]

Going beyond the production of CO, we investigate the further reduction of CO to hydrocarbons. Insights into the factors controlling the branching between different hydrocarbon species and hydrogen as a side product are scarce, precluding any rational design of selective catalysts. To remedy these shortcomings, we devise a platform allowing for the systematic study of the relevant reaction parameters. Exploiting non-aqueous electrolytes, we isolate activation-controlled kinetic data in electrochemical driving force, proton availability, solvent dielectric and reagent concentration. Our observations provide unique insight into the mechanism of hydrocarbon production on copper electrodes, demonstrating that the competition between H and CO for surface sites plays an important role in governing the product selectivity between methane, ethylene and hydrogen. These results establish a new paradigm in controlling hydrocarbon product selectivity in electrochemical fuel formation.[2]

References

[1] M. Schreier, F. Héroguel, L. Steier, S. Ahmad, J. S. Luterbacher, M. T. Mayer, J. Luo, M. Grätzel, *Nat. Energy* **2017**, *2*, 17087.

[2] M. Schreier, Y. Yoon, M. N. Jackson, Y. Surendranath, *Angew. Chem. Int. Ed.* **2018**, DOI 10.1002/anie.201806051.

[3] G. P. Lau, M. Schreier, D. Vasilyev, R. Scopelliti, M. Grätzel, P. J. Dyson, *J. Am. Chem. Soc.* **2016**, *138*, 7820–7823.

[4] M. Schreier, P. Gao, M. T. Mayer, J. Luo, T. Moehl, M. K. Nazeeruddin, S. D. Tilley, M. Grätzel, *Energy Env. Sci* **2015**, *8*, 855–861.

[5] M. Schreier, L. Curvat, F. Giordano, L. Steier, A. Abate, S. M. Zakeeruddin, J. Luo, M. T. Mayer, M. Grätzel, *Nat. Commun.* **2015**, *6*, 7326.

[6] M. Schreier, J. Luo, P. Gao, T. Moehl, M. T. Mayer, M. Grätzel, *J. Am. Chem. Soc.* **2016**, *138*, 1938–1946.

531445 Bimetallic Nanoporous Pd Alloys as CO Tolerant Electrocatalysts for the Electrohydrogenation of CO_2 to Formate

Abstract:

Electrochemical CO_2 reduction reaction (CO_2RR) has been widely investigated for conversion of water dissolved CO_2 into useful carbon based products e.g CO, formate/formic acid, alcohols etc. Depending upon the binding affinities of the different reaction intermediates, metals available for the CO_2RR show variability in product selectivity, overpotential requirements and sustained activity. Thus development of a suitable electrocatalyst with a high activity/faradaic efficiency (FE) for a single product is a challenge[1]. Among the simple 2 electron reduced products, formate/formic acid is gaining focus because of its high energy density and utility as a chemical precursor. Traditional formate producing metals e.g. Sn, In, Bi, Co etc. require high overpotentials (>700 mV), limiting their effectiveness. Recently, Pd, originally known to produce CO at overpotentials above 0.6 V RHE[2], has recently demonstrated formate FE's near unity at overpotentials below 0.2 V[3]. The gradual poisoning of Pd-based CO_2RR electrocatalysts through the slow evolution of CO, even at near unity formate FEs[4], results in catalyst deactivation and limits the viability of Pd in commercial applications.

Here, we report fabrication of free standing, core-shell, nanoporous bimetallic Pd (np-PdX, where X = Ag, Cu, Ni, Co) alloys that display a compositional dependent CO deactivation rate. The np-PdX electrocatalysts are made through electrochemical dealloying[5] of 4 binary $Pd_{15}X_{85}$ (X = Co, Ni, Cu, Ag) alloys. Electrochemical dealloying drives the formation of Pd skinned nanoporous architectures where the presence of less noble metal under the Pd shell alters the electronic structure of the Pd skin and consequently affects the CO adsorption strengths thereby changing the rate of catalyst deactivation. The np-PdX electrodes have a residual composition of ~ 30 atomic % of the less noble metal and pore sizes ranging from 5 – 10 nm. The tortuous pores give rise to roughness factors above 500 which generates high geometric formate partial current densities (> 40 mA/cm^2). Furthermore, the np-Pd electrodes are free standing and obviate the use of any binder and/or support which otherwise causes extra overpotential losses and morphological instability.

Upon performing extended electrolysis, the CO_2RR activity, formate FE and degree of deactivation of the np-PdX elecrodes are compared with a traditional Pd/C electrocatalyst having particle sizes of 5-10 nm. While all the np-PdX electrodes and Pd/C show above 90% formate FE within 200 mV overpotential, there is a significant variation in tolerance to CO poisoning as evidenced by the compositionally dependent time of deactivation. While np-PdAg and np-PdCu show faster deactivation compared to Pd/C, np-PdCo and np-PdNi avoid poisoning for much longer times. The results are explained on the basis of weakening of CO binding strengths due to electronic interactions of the alloying component that shifts the d-band center of Pd[6]. This is also shown by the trend of weighted average CO stripping peak potentials on Pd-skinned polycrystalline PdX (X = Co, Ni, Cu and Ag) alloys. Thus np-PdX electrocatalysts based on a suitable choice of alloying component, offer substantial suppression of CO poisoning, much higher formate geometric current densities and enhanced operational stability compared to traditional Pd based CO_2RR electrocatalysts.

References:

1. Kuhl, K., Cave, E., Abram, D., Jaramillo, T. New insights into the electrochemical reduction of carbon dioxide on metallic copper surfaces. *Energy Environ. Sci.* **5,** 7050–7059 (2012).
2. Gao, D. *et al.* Size-Dependent Electrocatalytic Reduction of CO_2 over Pd Nanoparticles. *J. Am. Chem. Soc.* **137,** 4288-4291 (2015).
3. Klinkova, A. *et al.* Rational Design of Efficient Palladium Catalysts for Electroreduction of Carbon Dioxide to Formate. *ACS Catal.* **6,** 8115–8120 (2016).
4. Min, X. & Kanan, M. W. Pd-Catalyzed Electrohydrogenation of Carbon Dioxide to Formate: High Mass Activity at Low Overpotential and Identification of the Deactivation Pathway. *J. Am. Chem. Soc.* **137,** 4701–4708 (2015).
5. Li, X. *et al.* Dealloying of Noble-Metal Alloy Nanoparticles. *Nano Lett.* **14,** 2569–2577 (2014).
6. Ou, L. Design of Pd-Based Bimetallic Catalysts for ORR: A DFT Calculation Study. *J. Chem.* **2015,** 1–11 (2015).

534365 Electrochemical Promotion of Catalysis: Non-Faradaic Effects of Applied Potential on CO_2 Hydrogenation and Ethylene Oxidation Reactions

Abstract:

Electrochemical promotion of catalysis is achieved in gas phase reactions over solid catalysts via the coordination of a dual-function working electrode/catalyst and catalytically inert counter and reference electrodes connected via a purely ionically conducting solid electrolyte (H^+ or O^{2-} conductor). Catalytic reaction rates of CO_2 hydrogenation and C_2H_4 oxidation can be reversibly modified by nearly two orders of magnitude via the application of a potential between the catalyst and reference electrodes. Faradaic efficiencies, defined by the ratio of the change in catalytic rate (dr) to the galvanostatically-defined rate of ionic species supplied to the catalyst surface (I/nF), are far in excess of unity. These non-Faradic processes result in changing selectivity of CO_2 reduction catalysis over Ru at 693 K by increasing CH_4 formation rates and concurrently reducing CO formation rates with the application of a positive potential, and positive potential application modifies C_2H_4 oxidation reaction rates at 693 K over Pt by over an order of magnitude with Faradaic efficiencies in excess of 50. Applied potentials result in the changing reactant equilibrium binding constants as modeled by Langmuir-Hinshelwood-type rate expressions. Potential application results in the back-spillover of ionic species from the solid electrolyte onto the ionic surface to form an electrically neutral effective double layer that can be quantified via analysis of galvanostatic transients in conjunction with C_2H_4 oxidation kinetic modeling. This work investigates and models the fundamental non-Fardaic effects of applied potential upon a catalytically active surface with regards to gas phase reactions via alteration of site densities and energetics.

535146 Nano- to Macro Scale Morphological Impacts on CO_2Â electroreduction Product Selectivity over Cu Catalysts

Abstract:

CO_2 electroreduction has recently attracted a great deal of attention as a method to produce highly valuable chemicals and fuels from the waste CO_2. Among the possible products generated in this process, hydrocarbons such as methane and ethylene are of interest owing to their wide applications. Cu is the only metal catalyst which can produce hydrocarbons in significant quantities. However, it needs a high overpotential and also produces other side products which are not as desirable. Among the factors affecting the product selectivity and efficiency of the Cu catalysts, the surface morphology plays a significant role[1-4]. Although it has been observed that increasing the roughness, which is a morphological parameter, is frequently associated with ethylene formation enhancement, the reason for this relationship is still under investigation[1, 4-5].

In this study, the effect of different nanoscale morphologies on the product selectivity was investigated. The Cu catalysts were synthesis by electropoishing and electrodeposition. We found a trend between electrochemical roughness and the faradaic efficiency (%FE) of methane and ethylene. By increasing the roughness from 1.6 to 7.8, the product selectivity changed from methane formation to ethylene formation. On the Cu/Cu 1.6 catalyst, we obtained 28%FE towards methane and 12%FE towards ethylene at -1.18V vs. RHE. In contrast, on the Cu/Cu 7.8 catalyst, 36%FE towards ethylene and 8%FE towards methane was observed at the same potential. Roughness follows the product selectivity frequently but not all time. We have found that electrochemical roughness can be similar on very different catalyst morphologies (Cu/Cu 1.6 and Cu/Cu 1.5) and result different product selectivities, illustrating more basic underlying effects are in play. We observed high selectivities toward ethylene on Cu rough catalysts that had higher current intensities on the surface of the catalyst. We have also observed that this current effect can be translated across length scale from nano to macro. We have found that by changing the electrode geometry from Cu flag shaped to Cu wire shaped, the product selectivity also can be switched between methane and ethylene. Moreover, it was observed that Cu wire had a higher catalytic activity compared to the Cu flag.

1. Tang, W.; Peterson, A. A.; Varela, A. S.; Jovanov, Z. P.; Bech, L.; Durand, W. J.; Dahl, S.; Norskov, J. K.; Chorkendorff, I., *Phys. Chem. Chem. Phys.* **2012,** *14* (1), 76-81.
2. Karaiskakis, A. N.; Biddinger, E. J., *Energy Technol.* **2017,** *5* (6), 901-910.
3. Reske, R.; Mistry, H.; Behafarid, F.; Roldan Cuenya, B.; Strasser, P., *Journal of the American Chemical Society* **2014,** *136* (19), 6978-6986.
4. Mistry, H.; Varela, A. S.; Bonifacio, C. S.; Zegkinoglou, I.; Sinev, I.; Choi, Y.-W.; Kisslinger, K.; Stach, E. A.; Yang, J. C.; Strasser, P.; Cuenya, B. R., *Nature Communications* **2016,** *7*, 12123.
5. Chen, C. S.; Handoko, A. D.; Wan, J. H.; Ma, L.; Ren, D.; Yeo, B. S., *Catalysis Science & Technology* **2015,** *5* (1), 161-168.

538172 Insights into the Electrocatalytic Conversion of CO_2 into CO, Ethylene, and Ethanol in Alkaline Media

Abstract:

Using CO_2 as a feedstock for the production of intermediates for chemical or fuel production such as formic acid, CO, ethylene, ethanol, and methanol is one of several approaches being explored to help reduce anthropogenic CO_2 emissions, while also reducing society's dependence on diminishing fossil fuel reserves. A range of active electrocatalysts for the selective reduction of CO_2 have been reported. We have explored both Ag-based and Au-based catalyst as well as micro-structured CuAg alloy films, respectively for the selective and energy efficient reduction of CO_2 to CO [1,2] and to ethylene and ethanol [3]. Partial current densities for CO production exceeding 150 mA/cm^2 at cell overpotentials less than 1V now can be achieved routinely at CO energetic efficiencies in the range of 50-60%. Similarly, high surface area CuAg alloy films exhibit Faradaic efficiency for C_2H_4 and C_2H_5OH production of 60% and 25%, respectively, at a cathode potential of just −0.7 V vs RHE and a total current density of 300 mA/cm^2, the highest levels of selectivity at high activity and low applied potential reported to date.

This talk will focus on various mechanistic insights of the CO_2 electroreduction process on these different catalysts, especially in the presence of alkaline electrolyte. For example, the onset cathode potentials, the kinetic isotope effect, and Tafel slopes indicate that the production of CO at low overpotentials in alkaline media is the result of a pH-independent rate-determining step (i.e., electron transfer) in contrast to a pH-dependent overall process. Furthermore, in situ Raman and electroanalysis studies suggest the origin of the high selectivity toward C2 products observed with CuAg alloy films to be a combined effect of the enhanced stabilization of the Cu_2O overlayer and the optimal availability of the CO intermediate due to the Ag incorporated in the alloy.

1. Gold Nanoparticles on Polymer-Wrapped Carbon Nanotubes: An Efficient and Selective Catalyst for the Electroreduction of CO_2, H.R.M. Jhong, C.E. Tornow, C. Kim, S. Verma, J. L. Oberst, P.S. Anderson, A.A. Gewirth, T. Fujigaya, N. Nakashima, P.J.A. Kenis, *ChemPhysChem*, **2017**, 18 (22), 3274-3279.
2. Insights into the Low Overpotential Electroreduction of CO_2 to CO on a Supported Gold Catalyst in an Alkaline Flow Electrolyzer, S. Verma, Y. Hamasaki, C. Kim, W. Huang, S. Lu, H.R.M. Jhong, A.A. Gewirth, T. Fujigaya, N. Nakashima, P.J.A. Kenis, *ACS Energy Lett.*, **2018**, 3, 193-198.
3. Nanoporous Copper–Silver Alloys by Additive-Controlled Electrodeposition for the Selective Electroreduction of CO_2 to Ethylene and Ethanol, T.T.H. Hoang, S. Verma, S. Ma, T.T. Fister, J. Timoshenko, A.I. Frenkel, P.J.A. Kenis, and A.A. Gewirth, *J. Am. Chem. Soc.*, **2018**, accepted.

538456 Insights on the Electrochemical Reduction of Carbon Dioxide Using Solid Oxide Electrolysis Cells

Abstract:

Extensive use of fossil fuels and consequential high levels of carbon dioxide (CO_2) emissions are major contemporary challenges.[1] The most direct way to mitigate this process is to activate reverse chemical pathways in which CO_2 is reduced into high-energy molecules.[2] The electrochemical reduction of CO_2 to CO has attracted increasing attention, since CO represents a valuable intermediate to the production of synthetic fuels using established processes, such as Fischer-Tropsch.[2, 3] One attractive approach to electrochemically reduce CO_2 to CO is the use of solid oxide electrolysis cells (SOECs)[4]. SOECs are solid-state electrochemical systems that, in principle, can facilitate the electrochemical reduction of CO_2 to CO with potentially very high rates due to favorable kinetics at high operating temperatures[2, 4]. While electrochemical reduction of CO_2 using SOECs offers a great deal of promise, these systems are still limited by the high activation overpotential losses induced by the sluggish CO_2 reduction kinetics at the cathode.[1, 4] Moreover, the current knowledge on the electrochemical reduction of CO_2 using SOECs is mainly limited to proof-of-concept experiments, demonstrating the viability of its operation. Lack in understanding of the governing mechanism for electrochemical reduction of CO_2 on SOECs, as well as, the catalytic role played by different cathode components (i.e., metal and the mixed metal oxides) make the optimization of the SOEC cathode activity very challenging. In the present work, we combine experimental and theoretical techniques to understand the chemical/electrochemical steps that govern CO_2 electrolysis on metals under SOEC conditions. Controlled electrochemical studies are used to investigate the role played by the metal and the mixed metal oxide cathode components toward the electrochemical reduction of CO_2. Structure/performance relations are developed in order to identify optimum cathode compositions for this process.

REFERENCES

[1] S.D. Ebbesen, S.H. Jensen, A. Hauch, M.B. Mogensen, High temperature electrolysis in alkaline cells, solid proton conducting cells, and solid oxide cells, Chemical reviews, 114 (2014) 10697-10734.

[2] X.-K. Gu, J.S.A. Carneiro, E. Nikolla, Heterogeneous electrocatalysts for CO2 reduction, in: Catalysis, 2017, pp. 94-121.

[3] K. Xie, Y. Zhang, G. Meng, J.T. Irvine, Direct synthesis of methane from CO 2/H2O in an oxygen-ion conducting solid oxide electrolyser, Energy & Environmental Science, 4 (2011) 2218-2222.

[4] X.-K. Gu, J.S. Carneiro, E. Nikolla, First-Principles Study of High Temperature CO2 Electrolysis on Transition Metal Electrocatalysts, Industrial & Engineering Chemistry Research, 56 (2017) 6155-6163.

540093 Revisiting Efficiency Limits of Solar Thermochemical Fuel Production By Non-Stoichiometric Ceria-Based Redox Cycling

Abstract:

Synthesis gas (syngas) production via non-stoichiometric ceria-based solar thermochemical redox cycles has been investigated in aspects ranging from materials thermodynamics and heterogeneous reaction kinetics to transport phenomena in porous ceria structures to solar thermochemical reactor engineering.[1,2] This technology promises relatively high theoretical solar-to-fuel efficiency. However, thermodynamic studies found in the literature employ inconsistent modelling assumptions, resulting in reported water splitting efficiency values varying broadly, from 0.04% to 40%.[3,4]

In the present study, we critically review thermodynamic analyses of previous pertinent studies. We focus on the thermodynamic description of the gas–solid flow configuration in the reduction and oxidation reactors. We present a new description leading to meaningful predictions of sweep gas and oxidizer demand based on conservation of mass and the Gibbs criterion. Further, we report on a novel optimization approach developed to accurately predict the efficiency limit at a system level for ceria-based metal oxides. We calculate a solar-to-fuel water splitting efficiency close to 8% for a system making use of a ceria-based metal oxide, temperature swings, pressure swings realized by inert sweep gas, gas–gas heat exchangers, and state-of-the-art gas separation technologies.

1. Bader, R. & Lipiński, W. in *Advances in Concentrating Solar Thermal Research and Technology* (eds. Blanco, M. J. & Santigosa, L. R.) 403–459 (Woodhead Publishing, 2016). doi:10.1016/B978-0-08-100516-3.00018-6
2. Wheeler, V. M. *et al.* Modelling of solar thermochemical reaction systems. *Sol. Energy* **156,** 149–168 (2017).
3. Ganzoury, M. A., Fateen, S.-E. K., El Sheltawy, S. T., Radwan, A. M. & Allam, N. K. Thermodynamic and efficiency analysis of solar thermochemical water splitting using Ce–Zr mixtures. *Sol. Energy* **135,** 154–162 (2016).
4. Krenzke, P. T. & Davidson, J. H. On the efficiency of solar H and CO production via the thermochemical cerium oxide redox cycle: The option of inert-swept reduction. *Energy & Fuels* **29,** 1045–1054 (2015).

540033 Experimental Framework for Evaluation of the Thermodynamic and Kinetic Parameters of Metal-Oxides for Solar Thermochemical Fuel Production

Abstract:

Experimental Framework for Evaluation of the Thermodynamic and Kinetic Parameters of Metal-Oxides for Solar Thermochemical Fuel Production

Richard J. Carrillo[1], Jonathan Scheffe[1]

[1]Department of Mechanical & Aerospace Engineering, University of Florida, Gainesville, USA 32611

Keywords: solar, thermochemical, hydrogen, fuel

The two-step metal oxide redox cycle is a promising and thermodynamically attractive means of solar fuel production. Experimental characterization of the metal oxides used in these cycles is often performed under unrealistic conditions (e.g. using excess oxidant) or with non-ideal control of reactant gases (e.g. TGA), making comparison of the performance of candidate materials difficult. To address this, we have developed a high-temperature tubular reactor in which thermodynamic and kinetic experiments can be readily performed that is capable of operating at temperatures up to 1873 K, total pressures ranging from ultimate vacuum to ambient, and oxygen partial pressures as low as about 10^{-30} atm (precisely controlled using an H_2/H_2O environment). The system was validated by performing isothermal relaxation experiments (wherein the sample environment was rapidly changed by altering the input vapor concentration) using ceria, and comparing measured oxygen nonstoichiometries with the literature; see Figure 1. For this presentation, we will discuss the utility of this reactor configuration and experimental scheme for producing accurate and repeatable measurements of the oxygen exchange capacities, thermodynamic properties (namely partial molar enthalpy, entropy, and Gibbs free energy changes), and kinetics of candidate redox materials under a wide range of conditions akin to those expected in solar thermochemical reactors.

Figure 1. Measured nonstoichiometry of pure CeO2 versus reactor temperature and oxygen partial pressure compared to measurements by Panlener et al. (1975) and model predictions by Tuller and Nowick (1979).

538052 The Effects of the Paired Charge Compensating Dopant Identity in Ceria for Solar Thermochemical H_2O and CO_2 Splitting

Abstract:

Solar thermochemical water and carbon dioxide splitting provide a pathway to high efficiency solar fuels. To-date, ceria remains the benchmark material due to its stability and rapid kinetics. Of single element dopants, only Hf- and Zr-doping has shown increased performance over undoped ceria. Recently, group IIIA/VA paired charge compensating dopants (PCCD) have provided a new pathway for altering the redox properties of ceria within an energy range commensurate with thermochemical cycles. Here we expand upon the initial PCCD strategy, expanding into new dopant combinations, using use density functional theory calculations. We show that the PCCD strategy offers a large opportunity for increasing the range of available reduction energies available in ceria and provide a fundamental understanding for the found behavior. Although these materials are investigated in light of solar thermochemistry, their behavior is of potential interest to the metal oxide catalytic community as a whole.

533785 Computationally Accelerated Discovery and Experimental Demonstration of Materials for Solar Thermochemical Hydrogen Production

Abstract:

While solar energy is the most abundant renewable energy resource, the capture, storage, and distribution of it remains a challenge. Solar thermochemical hydrogen production (STCH) provides a promising route for efficient utilization of this disperse resource since it allows for use of the entire solar spectrum to convert water to an energy dense fuel, H_2. However, despite a significant number of materials having been examined, an optimal redox material to drive this process has yet to be developed. In order to be viable for economic hydrogen production, materials must for have high hydrogen productivity, fast reduction and oxidation kinetics, low thermal reduction temperatures, and long term stability and reactivity. In this work we utilize a computationally accelerated, accurate, and experimentally validated materials-by-design approach involving *ab initio* and machine-learned models to design and demonstrate durable materials with improved thermodynamic and kinetic properties for STCH.

The thermodynamics of new materials are screened for stability, oxygen vacancy formation energy, and extent of reduction using machine learned models and DFT calculations. The effect of charged defects and their associated large electronic entropy on the predicted STCH behavior is assessed and compared to traditional charged-neutral screening methods. The effect of temperature on the parameters including crystal structure and cation disorder and their relationship to predicted STCH ability will be discussed. Compositional (doping) control is utilized to further optimize materialsâ€™ thermodynamic properties.

Thermogravimetric analysis (TGA) experiments are used to quantify the thermodynamic properties of new materials which provide a direct metric for comparison to computational results and help identify optimal operating conditions. Hydrogen production and kinetic parameters of new materials are measured during redox cycling in a stagnation flow reactor (SFR) and are provide feedback and validation to computational results. The long-term stability of new materials is critical to producing hydrogen economically. The stability of spinel particles is demonstrated over 200 redox cycles in the SFR and during on-sun operation at the high flux solar furnace at the National Renewable Energy Laboratory.

532291 Examining the Solar-to-Fuel Efficiency of Ceria and Perovskite Thermochemical Redox Cycles for Splitting H_2O and CO_2

Abstract:

Ceria and perovskite based materials have attracted significant attention for use in solar thermochemical H2O and CO2 splitting cycles. Ceria based materials, the current gold standard, tend to favor the oxidation reaction while perovskite materials tend to favored the reduction reaction. To date, most materials comparisons have focused on fuel production capacity per cycle and have neglected the effects of the plant and reactions on the overall solar to fuel conversion efficiency. Therefore, a thermodynamic model of a solar thermochemical gas (CO2/H2O) splitting plant was constructed and used to evaluate the performance of six candidate redox materials (two ceria and four perovskite materials) under an array of operation conditions. The values obtained for the solar-to-fuel energy conversion efficiency are higher in relative order Zr-doped CeO_2 > undoped CeO_2 > $La_{0.6}Ca_{0.4}MnO_3$ > $La_{0.6}Ca_{0.4}Mn_{0.6}Al_{0.4}O_3$ > $La_{0.6}Sr_{0.4}MnO_3$ > $La_{0.6}Sr_{0.4}Mn_{0.6}Al_{0.4}O_3$. This ordering is attributed to their relative reducibility and re-oxidizability, where the doped and undoped ceria, that favor oxidation, outperform perovskites, that favor reduction and therefore require high flowrates of excess H_2O and CO_2 during re-oxidation. Solids-solid heat recuperation during the temperature swing between the redox steps is crucial, particularly for ceria because of its low specific oxygen exchange capacity per mole and cycle. Conversely, the efficiencies of the perovskites are more dependent on gas-gas heat recuperation due to the massive excess of H_2O/CO_2. Redox material thermodynamics and plant/reactor performance are closely coupled, and must be considered together to maximize efficiency. Overall, we find that Zr-CeO_2 is the most promising redox material, while perovskites which seem promising due to high H_2/CO production capacities under large H_2O/CO_2 flow rates, perform poorly from an efficiency perspective due to the high heating duties, especially for steam.

533280 Concentrated-Light Aging Techniques for High-Temperature and Solar-Energy Materials: Preliminary Results

Abstract:

Energy supply may be the biggest economic issue that present-day society is facing. Although, most of the (fossil) energy resources currently exploited will not be depleting "soon", humanity and environment are calling for a sustainable model. Thus, green buildings and renewable energies are becoming worthwhile solutions [1]. Among them, solar thermal collectors [2], solar thermal electricity (STE) [3] seem to play a decisive role in the renewable energy mix, apart from photovoltaic technologies (PV) [4]. In any case, all technologies demand for more efficient [5] and durable materials [6], while the expected lifetime of most related components should exceed 20 years. In this context, state agencies declared materials to be a key enabling technology to help society lead into a resource-efficient, low carbon economy and highlighted that the absence of more complete testing methods, e.g. accelerated aging , inhibit and delay further improvements [7]. Aging generally refers to the degradation of given abilities over a certain period of time. These abilities depend on the technology and cover a wide range. For solar absorbers/receivers are absorption of solar radiation and heat transfer [8], for solar collectors/reflectors is reflectivity of solar radiation [6], and for photovoltaics is the optoelectric conversion [9]. Meeting and assuring of lifetime goals for all components is important for the success of solar energy technologies and for its competitiveness with other energy production technologies, especially with conventional ones that have already proved their durability and resilience.

In accelerated aging studies (accelerated tests-AT), generally, information from experiments at high levels of one or more accelerating variables (e.g. temperature, use rate or radiation) is extrapolated, through a physically reasonable statistical model, to obtain estimates of life or long-term performance at lower, normal levels of the accelerating variable(s) [10]. Indoor ATs for light induced degradation (photodegradation) approximate natural sunlight using, initially, Carbon-arc (since 1918) and, later, Xenon-arc (since 1960') light sources [11]. Concentrated sunlight devices have been developed as part of the outdoor ATs (weathering facilities, e.g. EMMAQUA [12]). Although, concentrated light is been used in outdoor ATs since 1980' [12] as a way to accelerate photodegradation by five to ten times, it was only recently introduced at higher rates. Organic photovoltaic materials (OPV) are extremely sensitive to light and especially UV, so they were among the first materials to be tested under concentrated light; first up to 27 suns (i.e. 27x the normal sunlight radiation) Tromholt *et al.* [20] and later up to 10,000 suns Katz *et al.* [21]. Note that, AT acceleration rates depend on the number of suns (also called concentration ratio) but it not exactly proportional. Recently, the combined effect of both temperature and radiation, acceleration variables, on OPV was also studied [22]. However, all these studies were conducted using concentrated sunlight that limits the operational time and control over the radiation, while some of the observed differences could be attributed to processes occurring during the dark periods [22]. Concerning other materials, Boubault *et al.* [8] used, again, homogenized concentrated sunlight to investigate an AT for a solar absorber material (Inconel625 coated with anon-selective paint coating Pyromark 2500) but by controlling only the total radiation and low concentration ratios, relative to the standard operation of solar absorbers. Cotfas *et al.* [9] presented an AT for photovoltaic cells (multijunction Emcore and monocrystalline) using concentrated artificial light but with no homogenization, thus, very poor control on the radiation levels. Most recently, our group displayed the thermal and optical evaluation of advanced solar receiver/reflector materials (MAX phase ceramics [23]) using a well-controlled environment (both temperature and radiation) under homogenized, artificial, and concentrated light [24].

The current problem of high-temperature and solar energy technologies, that our work deals with, is how to make up the scarcity of test procedures for performance evaluation and prediction-of-service-life of new materials. Following a rigorous analysis of the state-of-the-art, some of the most important limitations of current ATs, for solar energy materials, are the omission of the service-life light induced degradation and their low acceleration rates for this type of degradation, whenever included. Moreover, there is a lack of high-temperature AT for solar energy materials. (e.g. absorbers) and of facilities able to provide controlled AT environment using concentrating light. A situation even more problematic for research stage materials that typically are in need of faster screening, in other words ATs with higher acceleration rates. Here, we investigate the hypothesis that an accelerated aging technique based on concentrated light can address the aforementioned technology gaps.

References:

1. Solangi, K.H., et al., *A review on global solar energy policy.* Renewable and Sustainable Energy Reviews, 2011. **15**(4): p. 2149-2163.
2. Amri, A., et al., *Developments in the synthesis of flat plate solar selective absorber materials via sol-gel methods: A review.* Renewable and Sustainable Energy Reviews, 2014. **36**: p. 316-328.
3. Mills, D., *Advances in solar thermal electricity technology.* Solar Energy, 2004. **76**(1-3): p. 19-31.
4. Jäger-Waldau, A., *Photovoltaics and renewable energies in Europe.* Renewable and Sustainable Energy Reviews, 2007. **11**(7): p. 1414-1437.
5. Asim, N., et al., *A review on the role of materials science in solar cells.* Renewable and Sustainable Energy Reviews, 2012. **16**(8): p. 5834-5847.
6. García-Segura, A., et al., *Durability studies of solar reflectors: A review.* Renewable and Sustainable Energy Reviews, 2016. **62**(Supplement C): p. 453-467.
7. EC *Commission Staff Working Paper SEC(2011) 1609 Final. Materials Roadmap Enabling Low-carbon Energy Technologies.* European Comission: 2011.
8. Boubault, A., et al., *Aging of solar absorber materials under highly concentrated solar fluxes.* Solar Energy Materials and Solar Cells, 2014. **123**: p. 211-219.
9. Cotfas, D.T., et al. *Ageing of photovoltaic cells under Concentrated Light.* in *2015 Intl Aegean Conference on Electrical Machines & Power Electronics (ACEMP), 2015 Intl Conference on Optimization of Electrical & Electronic Equipment (OPTIM) & 2015 Intl Symposium on Advanced Electromechanical Motion Systems (ELECTROMOTION)*, 2015. 2-4 Sept. 2015.
10. Escobar, L.A. and W.Q. Meeker, *A review of accelerated test models.* Statistical Science, 2006. **21**(4): p. 552-577.
11. Scott, J., *Does correlation exist between accelerated and conventional outdoor exposures ?* Pigment & Resin Technology, 1980. **9**(2): p. 15-18.
12. Ellinger, M.L., *Correlation of weathering results.* Journal of Coatings Technology, 1977. **49**(627): p. 44-51.
13. Köhl, M., et al., *Durability of polymeric glazing materials for solar applications.* Solar Energy, 2005. **79**(6): p. 618-623.
14. Arndt, T., S. Richter, and M. Pasierb. *Accelerated laboratory weathering of acrylic lens materials.* in *AIP Conference Proceedings*, 2013.
15. Ott, T., et al. *Accelerated aging and contact degradation of CIGS solar cells.* in *2012 IEEE 38th Photovoltaic Specialists Conference (PVSC) PART 2*, 2012. 3-8 June 2012.
16. Brunold, S., et al., *Accelerated life testing of solar absorber coatings:: Testing procedure and results.* Solar Energy, 2000. **68**(4): p. 313-323.
17. Wette, J., et al., *Comparison of Degradation on Aluminum Reflectors for Solar Collectors due to Outdoor Exposure and Accelerated Aging.* Energies, 2016. **9**(11): p. 916.
18. Sharma, V. and S.S. Chandel, *Performance and degradation analysis for long term reliability of solar photovoltaic systems: A review.* Renewable and Sustainable Energy Reviews, 2013. **27**: p. 753-767.
19. Zhang, K., et al., *A review on thermal stability and high temperature induced ageing mechanisms of solar absorber coatings.* Renewable and Sustainable Energy Reviews, 2017. **67**(Supplement C): p.

1282-1299.
20. Tromholt, T., et al., *Effects of concentrated sunlight on organic photovoltaics.* Applied Physics Letters, 2010. **96**(7): p. 073501.
21. Katz, E.A., et al. *Accelerated stability testing of organic photovoltaics using concentrated sunlight*. in *2012 38th IEEE Photovoltaic Specialists Conference*, 2012. 3-8 June 2012.
22. Visoly-Fisher, I., et al., *Concentrated sunlight for accelerated stability testing of organic photovoltaic materials: towards decoupling light intensity and temperature.* Solar Energy Materials and Solar Cells, 2015. **134**(Supplement C): p. 99-107.

534701 Nano-Structured Ceramic ALD Coatings to Stabilize SiC Against Oxidation in High Temperature Steam Solar Thermal Water Splitting Applications

Abstract:

Silicon Carbide (SiC) is an ideal material for many high-temperature applications due to its resistance to thermal shock and high thermal conductivity. However, SiC degrades in water-rich environments limiting its applications in extreme oxidative environments such as combustion engines, heat exchangers, and high temperature solar thermal reactor materials. Current deposition methods for environmental barrier coating (EBCs) such as plasma spraying and aerosol spraying generate micron-scale films with inherent grain boundaries and cracks. Atomic layer deposition (ALD) generates nano-scale films that are dense, crack-free, and chemically bonded to the surface without the use of excess precursor. We have found that the application of ALD EBCs significantly improves the oxidation resistance of SiC in extreme environments with a film four orders of magnitude thinner than standard applications. We have grown ALD layers of our desired materials on high surface area SiC particles using a fluidized bed particle ALD reactor in order to investigate the chemistry of steam oxidation. Using thermogravimetric analysis, it has been shown that these coatings improve the oxidation resistance of SiC by up to 2/3 at 1000°C. Isothermal kinetic analysis was used to compare the kinetic effects of the ultra-thin coatings, and based on derived rate expressions, a 10 nm film is predicted to enhance the lifetime of SiC parts by 3X. We have successfully demonstrated the use of ALD coated SiC solar thermal reaction tubes on-sun at NRELâ€™s high flux solar furnace.

539749 Integrating Non-Precious Metal H_2 Evolution Catalysts into Water Electrolyzers and Photoelectrochemical Water-Splitting Devices

Abstract:

The past decade has produced significant advances in the development o f non-precious metal catalysts for electrochemical H2 evolution. This paper will describe recent efforts in this area, and importantly, translation to water electrolyzers and devices for photoelectrochemical water-splitting.

535226 When Electrocatalysis Matters and When It Does Not: Unexpected Observations in Water Electrolysis and Flow Battery Energy Storage

Abstract:

This presentation will summarize ongoing projects in our lab on applications of electrocatalysis in hydrogen production and flow battery energy storage. We are working to advance these technologies through a balanced effort in scientific inquiry (e.g., synthetic methods, structure-function relationships) and engineering design (e.g., device specification, performance metrics). A persistent theme in our work thus far has been ambitious experimental targets underpinned by naïve assumptions, where the process of stepping back to re-evaluate our assumptions has nevertheless led to intriguing new conclusions.

In the area of water electrolysis, we set out to improve on the performance and stability of Ni–Mo alloy hydrogen evolving electrocatalysts with an eye toward their use in alkaline electrolyzers. Our work focused initially on understanding composition and reaction mechanisms in nanostructured Ni–Mo alloys while we also worked to develop catalyst ink formulations that would be compatible with fabrication of membrane electrode assemblies. Ironically, the latter studies led to unexpected fundamental insights, where observed reaction rates were not limited by kinetics or mass transfer, but rather by interfacial electronic conductivity even in this nominally metallic catalyst material. Incorporating carbon black mitigated the conductivity limitation, resulting in markedly improved mass-specific activity.

In the area of flow battery energy storage, our initial efforts to pursue engineering design specifications for integrated solar-driven battery systems were quickly hampered by the remarkable lack of consensus on intrinsic rates of interfacial electron transfer for known flow battery redox couples. Thus, we have now developed analytical protocols (adapted from the electrolysis and fuel cell literature) for benchmarking the kinetics of flow battery electrode-electrolyte combinations. Results to date suggest that electrocatalysis plays a vital role in dictating the overpotential performance of essentially all known flow battery active materials—even those based on apparently straightforward 1-electron transfer reactions. We are working to modify these electroanalytical protocols for direct deployment in operational flow battery devices.

530509 Understanding the Role of Adsorbed Hydroxide in Reversible Hydrogen Reactions

Abstract:

The kinetics of HER/HOR on Pt and other catalysts is facile at low pH, however, reaction rates are found to be several orders of magnitude slower in base. The explanation for this unusual pH dependence has been a matter of debate within the scientific community. Traditionally, hydrogen binding energy (HBE) has been used as the lone activity descriptor, independent of electrolyte pH [1]. This alone, however, is not sufficient to describe the pH dependence of the HER/HOR activity on metals other than Pt. Markovic et al. have proposed that hydroxide adsorption is necessary to facilitate water recombination/dissociation in base [2].

In this work, our aim is to determine if adsorbed hydroxide plays an active or spectator role in HER/HOR kinetics combining electroanalytical techniques with microkinetic modeling. We use expressions for time-dependent current and voltage as a function of hydrogen and hydroxide binding strength for the indirect (hydroxide-mediated) and direct (hydroxide-as-spectator) Volmer steps in base to simulate cyclic voltammograms. The simulated results are compared to experimental cyclic voltammograms of H-UPD on Pt(110). The free energy of hydroxide adsorption is controlled experimentally by the electrolyte cation via changes in cation solvation strength [3]. Varying the sweep rate results in greater peak separation that can be used to extract rate constants for the adsorption reaction via traditional electroanalysis [4].

Our computations show that at both slow and fast scan rates, either the direct or indirect mechanism can describe the peak potential splitting and peak current. However, hydroxide binding strength affects the mechanisms differently. In the indirect mechanism, stronger hydroxide binding reduces overpotential and, therefore, peak splitting. In the direct mechanism, stronger hydroxide binding decreases available surface sites, leading to lower exchange current densities and greater overpotential. Comparison with experiments show that from 0.1M KOH to 0.1M LiOH the peak potential location decreases while peak potential splitting increases. This trend shows that stronger hydroxide binding has detrimental effects on H-UPD kinetics, and that the direct mechanism is therefore more likely to dominate observed behavior.

Our results strongly suggest that weakening OH binding strength should increase availability of sites for H adsorption. And therefore, oxophilicity is not an accurate descriptor for alkaline hydrogen electrocatalysts and other parameters, such as water orientation and non-covalent interactions, might be playing a greater role in the overall activity.

References:

(1) Durst et al. Energy Environ. Sci. 2014, 2255.

(2) Strmcnik et al, Nat. Chem. 2013, 1.

(3) Danilovic, N. et al. Electrocatalysis 3, 2012, 221.

(4) Angerstein-Kozlowska et al. J. Electroanal. Chem. Interfacial Electrochem. 1979, 1.

532712 Electrocatalysis at Buried Interfaces

Abstract:

Electrocatalysis underlies many emerging clean energy technologies, such as fuel cells and electrolyzers. However, continued improvements in electrocatalyst activity, durability, and selectivity must be achieved in order for these technologies to make meaningful contributions to a sustainable energy future. In this presentation, I will describe unconventional electrocatalyst architectures for which metallic electrocatalysts are encapsulated by ultra-thin overlayers made of permeable oxides. These oxide layers are synthesized with a room temperature process and deposited as uniform and continuous layers onto thin films of Pt, one of the most commonly used electrocatalyst materials. Through various electroanalytical measurements, we show that these oxide overlayers can be selectively permeable to certain electroactive species and thereby enable efficient and selective electrocatalysis at the buried interface between the oxide overlayer and active catalyst. For this reason, these electrocatalysts may be referred to as membrane-coated electrocatalysts (MCECs). By systematically varying the thickness of silicon oxide overlayers in MCECs, we analyze their transport properties and the influence of mass transfer on the performance of thin film hydrogen evolution electrocatalysts. This talk will also describe characterization of the electroactive buried interface and discuss challenges and opportunities for controlling reaction pathways through modification of the physical, chemical, and/or structural properties at the interface. Importantly, this work highlights the potential of oxide-encapsulated electrocatalysts to serve as a tunable platform for efficient and stable electrocatalysis and/or photoelectrocatalysis of a variety of electrochemical reactions.

References

[1] N. Y. Labrador, X. Li, Y. Liu, J. T. Koberstein, R. Wang, H. Tan, T. P. Moffat, and D. V. Esposito, Nano Letters, vol. 16, 6452-6459, 2016.

[2] D.V. Esposito, *ACS Catalysis,* 2018, vol. 8, pp 457â€“465.

[3] N. Y. Labrador, E. L. Songcuan, C. De Silva, Han Chen, Sophia Kurdziel, Ranjith K. Ramachandran, Christophe Detavernier, D.V. Esposito, *ACS Catalysis, vol. 8, 1767â€“1778,* 2018.

528647 Density-Functional-Theory Studies of Face-Centered-Cubic Tungsten Carbide and Pt Core-Shell Nanoparticles Catalysts for the Hydrogen Evolution Reaction

Abstract:

Platinum is among the best electrocatalysts for the hydrogen evolution reaction (HER)—a potentially fossil-free route for hydrogen production—but the scarcity and high cost of this noble metal catalyst makes it economically unviable at commercial scales. Reduction in Pt loading and overall catalyst cost, without sacrificing catalytic activity, can be achieved by the synthesis of highly stable core-shell nanoparticles of transition-metal carbides and Pt. A large body of prior theoretical work has focused on the hexagonal phase of WC (α-WC) as a support material for Pt monolayers (ML), while the face-centered-cubic phase of WC (β-WC) has attracted lesser attention. Motivated by the recent preparation of stable β-$Ti_xW_{1-x}C$@Pt core-shell nanoparticles, we employ density-functional theory (DFT) calculations to study and contrast the suitability of the α and β phases of WC as support (core) materials for Pt ML (shells). We examine the thermodynamic stability of Pt MLs on α- and β-WC surfaces, carefully accounting for the delicate balance between epitaxial mismatch strains and chemical bonding between cores and shells. Furthermore, we study the effects of alloying β-WC with Ti to modulate the stability and electronic structure of the core-shell structures. Based on our thermodynamic studies, we find that β-WC is a better support material for Pt MLs than α-WC and that alloying with Ti improves the stability of Pt/β-WC interfaces without adversely affecting the activity of the Pt shell. We compare the electronic structures of Pt MLs supported on α-WC and β-$Ti_xW_{1-x}C$ surfaces and compare their activities for HER using the hydrogen binding energy (HBE) as a descriptor of catalytic activity. Of the various cases studied, we find that 2ML Pt supported on β-WC (111) and β-$Ti_{12.5}W_{87.5}C$ (111) show higher HER activities than Pt (111) making these core-shell structures very promising candidates for HER electrocatalysts; considering the improved thermodynamic stability of β-$Ti_{12.5}W_{87.5}C$ over β-WC, the former is predicted to be a particularly suitable core material for Pt shells. Overall, our results provide detailed insight into experimental observations of the high catalytic activity of β-$Ti_xW_{1-x}C$@Pt core-shell nanoparticles.

535328 Kinetic Investigation of Nickel-Iron Layered Double Hydroxide for Hydrogen Evolution in an Alkaline Electrolyte

Abstract:

Nickel-iron layered double hydroxide (NiFe LDH) and other transition metal layered materials have recently gained attention for their simple preparation methods, earth abundance, low cost, and most importantly, high activity toward hydrogen evolution (HER) and oxygen evolution (OER) reactions [1],[2]. In this work, we investigate the kinetics of an active NiFe LDH toward HER in an alkaline electrolyte. The mechanism of HER starts with electrochemical adsorption of a water molecule onto an active site on the cathode to produce the adsorbed species H (a step that is typically referred to as the Volmer step). The hydrogen gas can then be produced from two competing reaction paths: a chemical recombination of the adsorbed species H (Tafel step), and an electrochemical desorption step (Heyrovsky step).

Our NiFe LDH films were deposited on Ni foam substrates via a simple one-step hydrothermal synthesis route in an autoclave reactor maintained at a temperature of 155 C for 12 hr [1]. The films were then used as both OER and HER electrodes in a three-electrode electrochemical cell set up with 1 M KOH solution as an electrolyte.

The deposited film demonstrated an overpotential of 247 mV at a current density of 10 mA/cm^2 toward the HER. Linear sweep voltammetry (LSV) data were obtained at a scan rate of 1 mV/s between the cathodic potential range of -0.9 to -1.7 V vs Hg/HgO reference electrode. The calculated cathodic overpotential measurements were corrected for the voltage drop corresponding to the uncompensated solution resistance. A Tafel plot constructed from the data shows a Tafel slope of -38.4 mV/dec, which indicates that the HER mechanism is controlled by either the Herovsky or Tafel steps [3]. In addition, electrochemical impedance spectroscopy (EIS) measurements were taken at different potentials within the LSV potential range. The EIS data were then fitted to an Armstrong and Henderson equivalent electric circuit using both Z-view software and EIS Spectrum Analyser to obtain the equivalent circuit components.

Using a least squares method, a theoretical model relating current density to the electrochemical reaction rates [4] was fitted with the LSV experimental results (Figure 1 attached). The chemical reaction rate constants for the individual HER mechanism steps - Tafel, Volmer, and Herovsky - were obtained as a result of the fitting procedure. The resulting kinetic model shows a good agreement between theoretically calculated Faradaic resistance and experimental EIS results (Figure 2 attached).

The kinetics data revealed the Herovsky step as the rate determining step, which is in agreement of the calculated Tafel slope. Furthermore, the kinetics data show a dependence of the HER mechanism on the applied potential. For overpotential values between 0 and -0.4 V (corresponding to an applied potential between -0.9 to -1.6 V vs Hg/HgO), the HER mechanism is a sequence starting with the Volmer step, followed by the Tafel step. At higher cathodic overpotential values, the mechanism starts with a Volmer step followed by a rate controlling Herovsky step, with negligible contribution of the Tafel step. It was also observed that in higher cathodic overpotential region, the Tafel step reaction rate is independent of applied potential. This can be attributed to the surface coverage of the adsorbed intermediate reaching a constant high coverage value in this potential region.

Figure 1: Experimental and simulated results of current vs overpotential

Figure 2: Experimental and simulated results of Faradaic resistance vs overpotential

References:

[1] Lu, Z. et al. (2014). Three-dimensional NiFe layered double hydroxide film for high-efficiency oxygen evolution reaction. *Chemical Communications (Cambridge, England), 50*(49), 6479-82. doi:10.1039/c4cc01625d

[2] Luo, J. et al. (2014). Water photolysis at 12.3% efficiency via perovskite photovoltaics and Earth-abundant catalysts. *Science (New York, N.Y.), 345*(6204), 1593-6. doi:10.1126/science.1258307

[3] Zhu, Y., Liu, T., Li, L., Song, S., & Ding, R. (2018). Nickel-based electrodes as catalysts for hydrogen evolution reaction in alkaline media. *Ionics : International Journal of Ionicsthe Science and Technology of Ionic Motion, 24*(4), 1121-1127. doi:10.1007/s11581-017-2270-z

[4] Krstaji, N., Popovi, M., Grgur, B., Vojnovi, M., & Sepa, D. (2001). On the kinetics of the hydrogen evolution reaction on nickel in alkaline solution Part I. The mechanism. *J. of Electroanalytical Chemistry, 512*(1-2), 16-26.

537637 $BiVO_4/WO_3$ Photoanodes for Chloride Oxidation with Simultaneous H_2 Production

Abstract:

The most actively pursued application of photoelectrocatalytic fuel production is the decomposition of water to make hydrogen and oxygen (a.k.a. solar water splitting). Although the physical phenomena are reasonably well understood, the hurdle with solar water splitting is the development of a commercially a cost-effective system that can produce H_2 fuel with an overall energy efficiency of at least 10% or higher for a sustained period of time. One reason water splitting is expensive, is one half of the reaction is the 4 electron oxidation of water which is kinetically sluggish and results in the production of O_2, which has limited economic value. Electrooxidation of chloride ions which are increasingly abundant could provide an attractive, cost-effective option to water oxidation.

The focus of the current work is to carry out chloride oxidation in acidic pH (pH 1) using nanoporous bismuth vanadate ($BiVO_4$) as the light absorber, and use electrons provided from the oxidation step to generate hydrogen at the dark cathode. $BiVO_4$ has been widely regarded as a promising low-cost photoanode for solar to chemical energy conversion. However, most of the demonstrated work with $BiVO_4$ so-far has been limited to neutral or slightly basic pH, where $BiVO_4$ has been shown to be chemically and photoelectrochemically stable. For chloride oxidation, the pH of the electrolyte should be less than 3 to obtain high chlorine production yield. To address the above challenge, we report on a method to electrodeposit tens of nanometer thick amorphous tungsten oxide (WO_3) films on top of nanoporous crystalline $BiVO_4$ electrodes, and benchmark its activity for Cl^- oxidation.

525389 On the Combustion Synthesis and Characterization of $Ga_xZn_{1-x}O_yN_{1-y}$ for Water Splitting Applications

Abstract:

Utilization of solar energy is a necessary path forward to deal with the looming energy crisis and worldwide effects of fossil fuel usage. A key component of this utilization is how to store excess solar energy for use on demand. One such method is splitting water into hydrogen and oxygen, both of which can be stored and either burned as fuel or used as feedstock components in industrial processes to synthesize high-value materials. Metal oxynitrides have shown great potential as photocatalysts for overall water splitting applications. Incorporation of nitrogen into metal oxides narrows the band gap to enable absorption of light in the visible spectrum. The typical synthesis of a metal oxynitride, however, requires annealing temperatures at or above 900 ^oC and nitridation by flowing pure ammonia gas for anywhere from 1 to 110 hours. Such harsh synthesis conditions can limit the discovery of new candidate materials.

We have prepared an array of gallium zinc oxynitrides ($Ga_xZn_{1-x}O_yN_{1-y}$) as a model system using a combustion synthesis method. Gallium zinc oxynitrides can be synthesized in 30 minutes in an air atmosphere at temperatures ranging from 400-750 ^oC. XPS analysis reveals that nearly 50% nitrogen substitution in the metal oxide lattice can be achieved, dependent on the ratio of gallium and zinc nitrates used as precursors. UV-visible spectroscopy also demonstrates that the band gap of the $Ga_xZn_{1-x}O_yN_{1-y}$ materials can be controlled simply by altering the precursor ratio.

The as-synthesized products are photoactive under both UV-Visible and visible-only irradiation with and without hole scavengers present and without any co-catalysts. More importantly, photocurrent is measured in a 2-electrode setup with only pure water present. This presentation will discuss the physical and photoelectrochemical characterization of these materials to confirm successful synthesis, the current issues we are facing, paths forward to address these issues, and the ultimate goal of a high-throughput synthetic methodology.

539944 Theoretical and experimental investigation of effective solar mixed reforming for a less carbon intensive production of methanol

Abstract:

Mixed reforming of methane, i.e. replacing a fraction of the reactant steam by CO_2 has become a widely discussed option for the conversion of CO_2 into a useful product. However, this can only be beneficial, if the heat of reaction is provided by a heat source that does not produce CO_2 itself. For instance, concentrated solar irradiation is a suitable non CO_2 emitting heat source. Within the framework of the INDIREF project (Indirectly solar heated reforming for methanol production), the possibility of solar assisted mixed reforming for methanol production is investigated.

A theoretical study solely based on thermodynamic data was carried out to assess possible inlet ratios of $H_2O/CO_2/CH_4$ into the solar heated reforming reactor in dependence on reforming temperature and pressure. Several metrics where used to evaluate the results for the different configurations: (1) Absence of carbon formation, (2) sufficiently high methane (>85%) and CO_2 (>50%) conversion as well as a (3) suitable composition of the product syngas for methanol and Fischer-Tropsch synthesis. The results indicate that the criteria are only met within a very narrow range of $H_2O/CO_2/CH_4$ ratios. Furthermore it is shown that the maximum achievable CO_2/CH_4 ratio is 0.35, of which only 50 % are converted in the reactor. This shows the very limited capability of the mixed reforming reaction to convert CO_2 into a useful product. At the same time the study showed that low temperatures at the reformer inlet cause a regime that favors carbon formation in the reforming reactor. A new concept for later injection of the reactant CO_2 is proposed. The concept is validated by carrying out lab scale experiments in an electrically heated furnace on which results are presented.

In the next step, process for indirectly heated solar reforming with air as heat transfer medium is developed. The central innovation is the air heated reforming reactor, which allows for efficient utilization of the solar heat as well as effective reaction control based on the above mentioned findings. The developed process will be demonstrated in the large scale solar simulator (*synlight)* in Jülich, Germany in the end of 2018. Acknowledgements

This work was carried out with financial support from European Union through EFRE under contract EFRE-0800578 (INDIREF)

538016 Integrated Solar-Thermochemical Reactor/Gas Recuperator for Upgrading Light Hydrocarbons

Abstract:

Concerning the reduction of fossil-energy use from industrial processes and transportation, concentrated solar power from parabolic reflectors is well suited for driving highly endothermic reactions that valorize CO_2 and/or light hydrocarbons. In this study, an openable, lab-scale, directly-irradiated, solar-thermochemical millireactor â€“ integrated with gas recuperation architecture â€“ is designed, built and tested under simulated sun. Coupled on a single plate, steam reformation of methane and water gas shift reactions are explored as a route to hydrogen production. It is shown that integrated reactor/recuperator technology has potential to efficiently utilize solar input for higher value chemical production (>50%), as well as reduce the capital cost of current state of the art solar-hydrogen systems.

539929 Solar Thermochemical Hydrogen/Syngas Production from Methane and/or Biogas in the Presence of Non-stoichiometric Solid Oxide Carriers

Abstract:

Solar energy provides by far the largest of all carbon neutral-energy sources. Solar irradiation is, however, dilute, intermittent and non-equally distributed. These facts represent an enormous challenge for the development of solar processes. The conversion of solar energy into chemical energy carriers, i.e. solar fuels, which can be long term stored and/or transported, can alleviate some of these drawbacks. Solar fuels can be produced through thermochemical cycles based on redox metal pairs [1]. Non-stoichiometric oxides, such as CeO_2, doped CeO_2 and perovskites [2,3], involve oxygen release from their crystalline network. One of the main issues is the high temperature needed for the reduction step. Over 1400°C, lower overall efficiencies and important material problems may impede long-term cyclic operation. Carbothermal reduction, in the presence of carbon or methane [4], can bring down the temperatures needed for the 1st step. In the present work, the methanothermal reduction of CeO_2 and CeO_2-ZrO_2 has been considered. Their oxygen exchange capacity was tested at 700-1000°C, in order to produce syngas from methane and/or CH_4/CO_2 mixtures (simulating a typical biogas composition).

Equilibrium thermodynamic calculations were performed using HSC 5.0 Outokumpu software. Such thermodynamic calculations evidenced that more important deviations from stoichiometry (δ) can be obtained at the same temperature under carbothermal conditions, vis-à-vis CeO_2 reduction in inert atmosphere. For example, at 900°C, δ equals 0.239 for the carbothermal reduction in the presence of methane, whereas the δ reaches only 0.008 for the reduction of CeO_2 under inert atmosphere. Using 70% CH_4 – 30% CO_2 (simulated biogas) results in forecasted δ values amounting to 0.212 at 900°C, slightly lower than in the presence of pure CH_4. Moreover, a syngas of a certain quality is obtained since this first carbothermal reduction step. H_2/CO ratios range from 5.8 in the presence of CH_4, to 1.7 under 70% CH_4 – 30% CO_2. The overall syngas quality can be further adjusted when performing the second, $CeO_{2-\delta}$ oxidation step in the presence of steam, CO_2 or H_2O-CO_2. The theoretical solar-to-fuel efficiencies range from 0.47 to 0.57, just for the first carbothermal step in the presence of either CH_4 or 70% CH_4 – 30% CO_2, assuming that the oxidation of $CeO_{2-\delta}$ for the regeneration of CeO_2 can be simply performed using air. Taking into account the second H_2O and/or CO_2 splitting steps the cycle efficiencies increase respectively to 0.54 and 0.64.

Different CeO_2 and CeO_2-ZrO_2 oxides were used as oxygen carriers in carbothermal reduction experiments, performed in a fixed bed reactor heated by an electric furnace. The solid materials were oxidized at 500°C for 30 minutes under 20%O_2-Ar flow. CH_4-Ar or CH_4/CO_2-Ar were then fed to the reactor while the temperature was increased at 5°C/min until 900°C, followed by isothermal carbothermal reduction during 1 hour at 900°C. A mixture CO_2-Ar was finally fed to simulate the second (exothermal) step of the cycle, leading to the re-oxidation of the solid oxygen carrier. The results obtained confirm that H_2 and CO can be obtained in important amounts and at moderate temperature over CeO_2 and CeO_2-ZrO_2 solid oxygen carriers. Important differences were however observed when feeding CH_4-Ar and CH_4/CO_2-Ar. Under CH_4-Ar, δ reaches 1.47 at the end of heating ramp, corresponding to peak H_2 and CO concentrations of 3 and 1.25%, respectively. During subsequent isothermal carbothermal reduction at 900°C, H_2 and CO evolution decrease and stabilize at 1.3 and 0.5%. Upon 1-hour carbothermal reduction δ finally reaches a value of 2.18. This total amount of oxygen released from the solid oxygen carrier can be completely restored by feeding CO_2-Ar. CO evolution and the resulting oxygen balance point indeed to a complete regeneration of the O-stoichiometry in the original CeO_2. In the presence of CH_4/CO_2-Ar, δ constantly increases while heating, reaching values around 0.55 at 900°C. CO concentration reaches 3%, corresponding to H_2/CO ratios around 1. In the further isothermal step, δ does no longer increase and stabilizes at 0.57. When feeding CO_2-Ar, only a very small amount of the oxygen released from the non-stoichiometric oxide is restored into the solid, pointing to oxygen exchange occurring between the gas phase, i.e. CO_2, and the solid oxygen carrier, and leading to an almost continuous regeneration of its non-stoichiometry. It must be also noted here that no carbon formation was observed during these experiments. The experimental efficiencies calculated from the syngas yields remain still relatively low. Further improvements are needed in order to boost the oxygen release from the non-stoichiometric oxide.

[1] M.E. Galvez, P.G. Loutzenhiser, I. Hischier, A. Steinfeld, Energ. Fuels 22 (2008) 3544.

[2] J.R. Scheffe, D. Weibel, A. Steinfeld, Energ. Fuels 27 (2013) 4250.

[3] T. Cooper, J.R. Scheffe, M.E. Galvez, R. Jacot, G. Patzke, A. Steinfeld, Energ. Technol. 3 (2015) 1130.

[4] M.E. Galvez, A. Frei, F. Meier, A. Steinfeld, Ind. Eng. Chem. Res. 48 (2009) 528.

540088 Dynamic Performance of Fischer-Tropsch Liquid Fuel Production from Solar-Assisted Supercritical Water Gasification of Algae

Abstract:

The coupling of solar supercritical water gasification (SCWG) of algae with downstream Fischerâ€"Tropsch (FT) synthesis is a promising approach for renewable biofuel production. The accurate prediction of the economic performance highly depends on the transient behaviour of the system. However, the integration of solar-driven SCWG with the conventional FT slurry bubble column reactors, which take days or weeks to achieve steady-state operation, is quite challenging, due to the transient nature of the solar resource. Microtubular reactors, on the other hand, can achieve steady-state much faster than the conventional FT reactors.

This paper focuses on the effect of solar variability and evaluates the techno-economic feasibility of an integrated solar-SCWG and microtubular-FT process. For this purpose, the steady-state SCWG and FT processes are modelled in ASPEN Plus software. A fixed-size 50 MW_{th} con- centrated solar-thermal (CST) input is used to drive the endothermic reactions in the SCWG system. The output gases from the gasifier mainly contain CO2, CH4 and H2. Three alternative reforming technologies are evaluated in this research including Steam Methane Reforming (SMR), Autothermal Reforming (ATR), and Partial Oxidation/Dry Reforming (PO/DR) to convert CH4 to syngas. After the reforming process, the H2 to CO ratio is less than the optimal ratio (2.1) for downstream FT and some make-up hydrogen is provided from low-temperature photovoltaic water-splitting (PV-electrolysis). Syngas storage acts as the buffer between the SCWG and FT units.

The main dynamic parameters considered here are the ramp times for the FT reactor, the size of syngas storage and the solar multiple for the FT system. The off-design production rates of syngas and liquid products are determined from the steady-state ASPEN models as a function of DNI. The polynomials obtained from these variables are imported into a system model in OpenModelica. The case study developed here is based on one year of hourly weather data for Geraldton, Australia. The total capital investment of the integrated units is estimated using the nth plant scenario with a 30-year plant lifetime. The levelised cost of fuel (LCOF) is calculated as a function of the FT size, assuming a fixed size for CST. The sensitivity analysis suggests that the operating cost of algal and H2 (from PV-electrolysis) feedstocks are the dominant components of the LCOF. A single-objective optimization is performed to find the configuration with the lowest LCOF. This carbon-neutral fuel produced from algal biomass can be blended with fossil fuels to decrease the life-cycle greenhouse gas emissions. Further reduction in the LCOF is expected from technological improvements and cost reductions in PV-H2 technology, and using cheaper feedstocks.

537118 Solar Thermal Ethane Cracking

Abstract:

Concentrated solar thermal energy has the potential to provide process heat for chemical processing. To demonstrate the utility of concentrated solar, a directly-heated endothermic chemical process—ethane cracking—was carried out using simulated, concentrated sunlight. A miniaturized reactor suitable for the direct utilization of solar thermal energy for the purpose of ethane cracking was designed, built and tested. The effect of different process conditions (temperatures of 800 – 1000°C, steam-to-ethane ratios of 0 – 3.3, and mean residence times of 0.14 – 0.44s) on the system performance were investigated using a dilute mixture of ethane (3%) in nitrogen. Single-pass, ethane conversions as high as 69% were observed with high selectivity towards ethylene. This work demonstrates a new and promising approach to effective utilization of solar thermal energy for value added chemical production.

539943 Scope of zero-carbon metal production: State-of-the-art and future prospects

Abstract:

The rapidly growing infrastructure development around the world has led to a continually increasing demand for metal production. The primary energy requirement and green house gas (GHG) emissions associated with metallurgical extraction of structural metals, especially steel and aluminium, is quite high. Integration of renewable energy sources in the metallurgical industry is paramount to achieving the targets set out in the Paris Climate Agreement, 2015.

This work targets the iron and steel industry, which accounts for over 75% of the GHGemissions in the metal production sector. We focus on evaluating the various core reactor technologies for iron and steel production and discuss the opportunities for integration of renewables to eliminate fossil-fuel consumption and achieve reduction in GHG-emissions well beyond those achievable from efficiency improvements and modifications to the current direct reduced iron (DRI) and blast furnace (BF) routes. Relevant technologies evaluated include electrowinnning and pyro-electrolytic reduction of iron ore, solar-thermal integrated electrolysis and hydrogen-based steel making wherein the carbon-neutral H2 production is potentially achieved using PV-electrolysis, biomass gasification, photo-electrochemical route and solar thermal redox water-splitting. The primary energy requirement is calculated after implementation of a simple heat exchange network and pinch-analysis for each process. The potential for CO2-emission reduction for the various technological routes is presented and compared with the state-of-the-art. A case study relevant to the Australian context considering the economic impact and prospects of zero-carbon metal production is also presented.

Green Steam Initiative: Thermal performance of Cost Effective Parabolic Trough Collector System

***Deepankar Biswas*[1], *Vishwanath H. Dalvi* *[2], *Sudhir V. Panse*[3], *Jyeshtharaj B. Joshi* *[2,4], *Suresh P. Deshmukh*[1]**

[1] *Department of General Engineering, Institute of Chemical Technology, Mumbai*

[2] *Department of Chemical Engineering, Institute of Chemical Technology, Mumbai*

[3] *Department of Physics, Institute of Chemical Technology, Mumbai*

[4] *Homi Bhabha National Institute, Anushaktinagar, Mumbai*

*vh.dalvi@ictmumbai.edu.in, *jbjoshi@gmail.com

Abstract

It is an expressed objective of sustainable power source research to make solar power achieve value equality with power from fossil fuel based power plants. Solar thermal plants are capital escalated and don't profit firmly from economies of scale. Consequently, decreasing unit costs is the best way to a monetarily appealing innovation. Ordinary allegorical trough frameworks have confinements not minimum of which are the costly mirror and bolster necessities required to keep up high exactness in the optics. In an effort involving parabolic trough systems we have been able to develop an innovative, low cost, dismantlable and field assemblable solar collector reflector and receiver-housing costing 75 $/m2.

Keywords: Solar Energy; Parabolic Trough Collector (PTC); solar thermal system; solar steam generation, process heat, tracking system

Introduction

Parabolic trough collector (PTC) plants are the most developed systems of all commercially operating plants [1]. A typical parabolic trough consists of a linear parabolic mirror from which it reflects and concentrates the received solar energy onto a tube which is generally named as receiver tube positioned along the focal line [2], [3]. PTCs can only use the beam that come directly from the solar disk but not those that could be reflected from the surroundings that is Direct Normal Irradiance (DNI) [4].

Several researchers have generated calculation procedures for obtaining synthetic data on an hourly or daily basis. Knight et al. had carried out the models for generating hourly series of radiation and ambient temperature values, which aimed to replicate the statistics of long-term data with a year of generated data [5]. García-Barberena et al. evaluated the influence of an operational strategies on the performance of parabolic trough solar power plants with the aid of a computer program for the simulation of the energy behaviour of parabolic trough solar power plants [6]. Bonilla et al. established a dynamic simulator design and development of a direct steam generation parabolic trough solar thermal power plant [7]. Zhang et al. presented a baseline performance and economic evaluation of solar thermal systems based on the site test data and related references [8].

The performance of PTCs is important for the operation of the parabolic trough solar power plants location. In this paper a demonstration unit having an aperture area of $18m^2$ was set up and tested for steam generation. The proposed PTC system requires much lesser mirror area compared to conventional PTC design and require continuous automatic tracking for a daily operation.

PTC Module System design

The Parabolic trough solar collectors are designed to operate with tracking about only one axis. A tracking mechanism rotates the parabolic collector about an axis of rotation until the sun ray and the aperture plane is normal to the sun. The features expected from a PTSC module are that it should be easily transportable to the installation site, modular and scalable design. The installation time for installing a trough should be low. The parabolic trough has low costs motion systems for solar tracking but with an acceptable accuracy.

Table-1: PTC-Design key parameters

Feature/Description/Parameters	Value
Mounting, Sun tracking	N-S, E-W
Aperture width	3 meter
Length of single module	3 meter
Rim angle, ψ_{rim}	60°
Effective aperture area	18 m^2
Concentration ratio, CR	39.2
Absorber pipe diameter, Do	48.2 mm
Receiver absorptance, α	0.85
Receiver emittance, ε	0.5
Mirror reflectance, ρ	0.85
Glass plate transmittance, τ	0.9

Cosine factor	0.957

The dimensions of the PTC were based on the compact and modular design which can be easily transported, filed assemblable and dismantled. The input parameters in the collector design were aperture width of PTC and the rim angle as shown in the figure 1. The rim angle considered was 60° and the width of the aperture as 3 meter. Considering the above parameters and design reported in [4], the length of the parabola (Eqn 1), radius of the parabola (Eqn. 3), focal length of the parabola (Eqn. 2) and diameter of the receiver (Eqn. 4) were calculated.

$$f = W_a / \tan\left(\frac{\varphi}{2}\right) \tag{1}$$

$$r = \frac{2f}{1 + \cos(\varphi)} \tag{2}$$

$$S = \frac{H_p}{2}\{\sec\left(\frac{\varphi}{2}\right)\tan\left(\frac{\varphi}{2}\right) + \ln[\sec\left(\frac{\varphi}{2}\right) + \tan\left(\frac{\varphi}{2}\right)]\} \tag{3}$$

$$D = 2r\sin(\theta) \tag{4}$$

$$CR = \frac{W_a}{\Pi D} \tag{5}$$

Experimental setup and procedure

The PTC system was designed in a way that the focus of the aperture is at 1.6meter as shown in the schematic Figure 3. The design parameters of the system are listed in Table 1. The PTC system has continuous tracking so that the collector acceptance angle is always maintained at 0 degree. The receiver pipe size was 1.8" (outer diameter of 48.3mm) with CR of 39.2 as per Eq (5). The length of each unit is maintained at 3 meter so as to conveniently fix the mirror strips of 1.5 meter length. To reduce the end losses, it is necessary to connect as many units in series as possible. In this case, there are two units attached in series with the help of a connecting shaft. The parabolic blades were cut out from a Mild Steel sheet using the laser cutting operation. To check the accuracy of the cutting, the profile blades were then traced over the default template to check the exact profile of the curvature.

In order to reduce the cost of the receiver pipe support structure, the receiver pipe was held with the help of Mild Steel plates which acts as the intermediate support for the receiver pipe. For the requirement of the perfectly straight receiver pipe, maximum of 2 units of above design could be used to support a 6meter receiver pipe over the 18m2 area. Therefore, the final configuration of the parabolic trough unit was two PTC units attached in series to form a single larger unit with multiple supports for the pipe at equidistant spacing. Therefore the total land area used was approximately 18m2 (3m x 6m). This arrangement has ensured the maximum effective utilization of the piping (minimum pipe length requirement) and reduced the auxiliary length of the pipe that requires to be insulated.

The piping connections from the tank were completely insulated by the Rockwool insulation material and were enclosed with aluminium cladding. The insulation wool thickness was maintained at 50mm across the complete piping. The receiver pipe casing box consist of the circular receiver pipe and tank was insulated

Fig. 1. Elements of PTC system main unit.

Equations used for determination the PTC efficiency

Heat absorbed by the fluid = $\dot{m}.C_P.(T_{out} - T_{in})$ (1)

Solar energy incident on the collector = DNI x Ac (2)

Bulk mean temperature of the fluid = $\frac{T_{out} + T_{in}}{2}$ (3)

Thermal Efficiency of the system is = $\eta = (\dot{m}.C_P.\frac{(T_{out} - T_{in})}{DNI\ x\ A_c}$ (4)

Result and Discussion

Variation of Solar Intensity and Useful Heat Gain with respect to time

Fig. 2a and Fig. 2b represents the variation of the solar beam radiation and useful heat gain by re-circulating the fluid with respect to the time. The figure shows that the solar intensity is high throughout the period upto the noon and then starts decreasing till 15:00 Hrs. The recirculating fluid picks up the heat and it increases w.r.t to the time and then it starts decreasing by the end of the day. The peak value of the heat gain by recirculating the water was found very high near the solar noon and then it gradually decreases. Thus it is understood that the variation of the heat gain by the fluid varies with respect to the variation of the solar intensity over the experimental period.

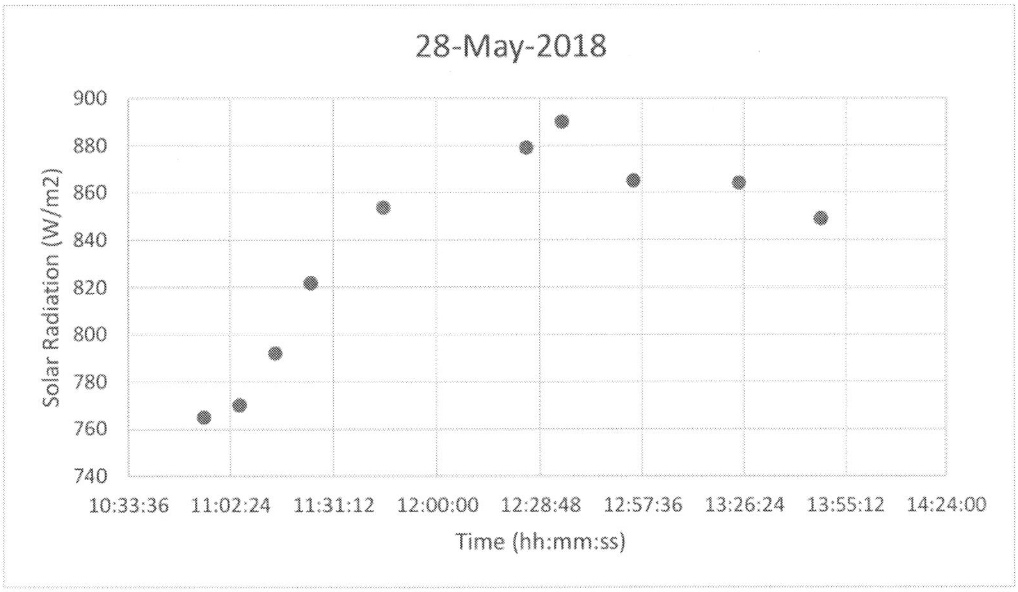

Fig. 2a. Variation of Solar radiation (W/m^2) with time (hr)

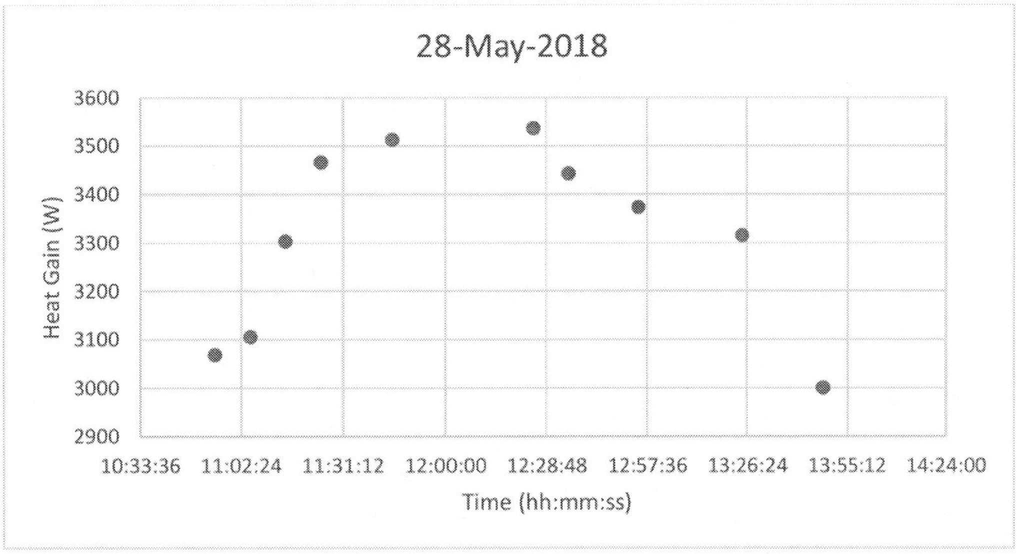

Fig. 2b. Variation of fluid heat gain (W) with time (Hr)

Variation of Outlet temperature with respect to time

Fig. 3 represents the variation of fluid outlet temperature with respect to time. It is observed that the outlet fluid temperature increases over the time period w.r.t to the increase in the solar intensity. The main reason is that the fluid is been recirculated from the storage tank and the fluid temperature of the tank increases over the entire heating period throughout the day. Thus it is clearly understood that the outlet temperature of the fluid follows the path of variation of the solar intensity.

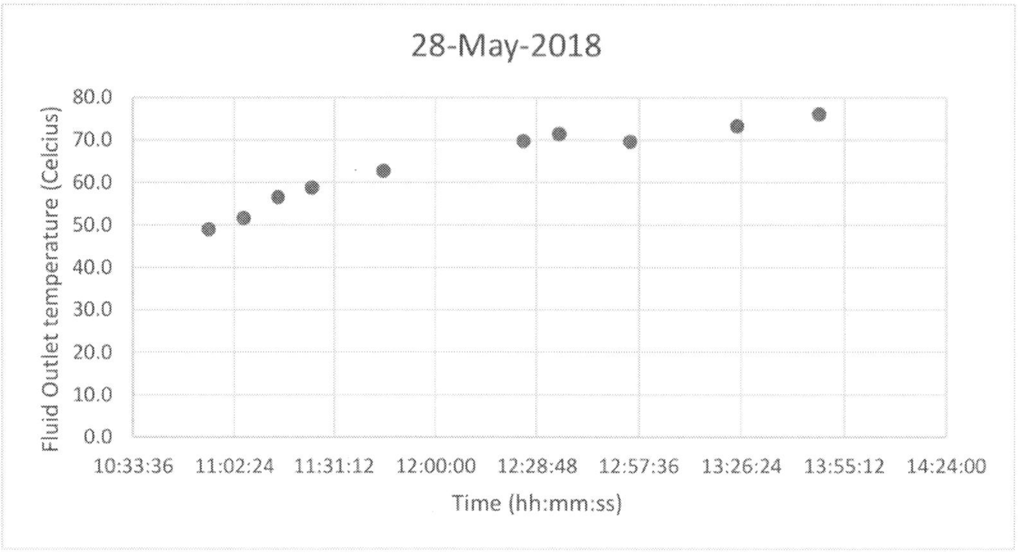

Fig. 3. Variation of fluid outlet temperature (°C) with time (Hr)

Variation of PTC efficiency with respect to time

Fig. 4 describes the variation in the efficiency of the system. The efficiency of the PTC increases rapidly and becomes stable during the solar noon. As the intensity of the sun starts decreasing over the time, the efficiency of the system drops. The peak value of the efficiency at the solar noon achieved is 30.16% at solar noon.

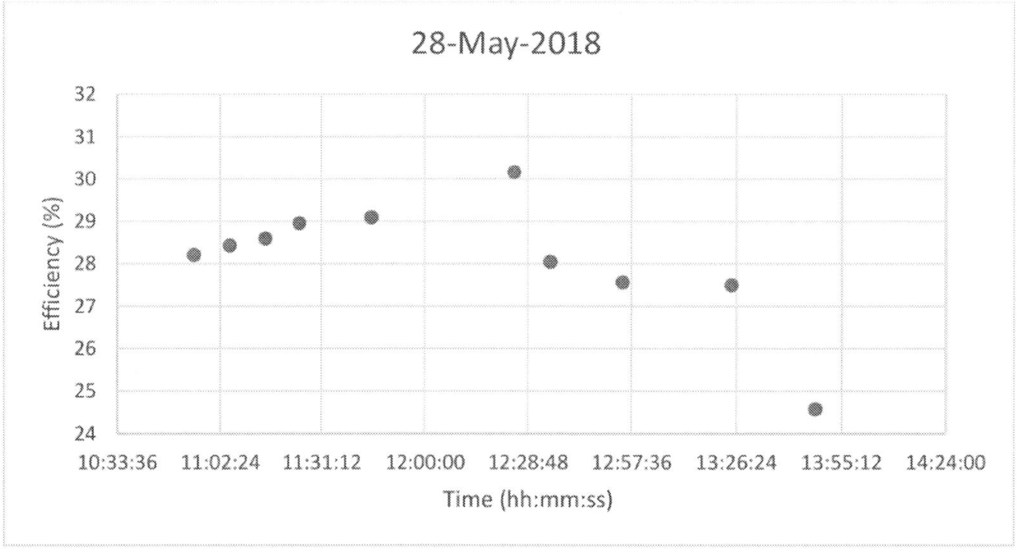

Fig. 4. Variation of efficiency with time (Hr)

Conclusion

From the experimental undertaken, it can be concluded that the useful heat gain and the efficiency of the trough increases till solar noon and then it starts to decrease and reach a minimum value. Also as the solar energy loss decreases, the heat gain in the receiver pipe increases. The fluid outlet temperature increases with respect to time.

Nomenclature

A_c	Effective aperture area of system (m^2)
C_p	Specific heat capacity of the system (J/kg-K)
$\dot{m}$	Mass flow rate of fluid (kg/sec)
PTC	Parabolic Trough Collector
T_i	Inlet temperature (°C)
T_{out}	Outlet Temperature (°C)
T_b	Bulk mean temperature
DNI	Direct Normal Irradiance (W/m2)
ψ_{rim}	Rim angle
η	Efficiency (%)

References

[1] H. L. Zhang, J. Baeyens, J. Degrève, and G. Cacères, "Concentrated solar power plants: Review and design methodology," *Renew. Sustain. Energy Rev.*, vol. 22, pp. 466–481, 2013.

[2] S. A. Kalogirou, "A detailed thermal model of a parabolic trough collector receiver," *Energy*, vol. 48, no. 1, pp. 298–306, 2012.

[3] H. Müller-Steinhagen and F. Trieb, "Concentrating Solar Power A Review of the Technology," *Ingenia Inf. QR Acad Eng*, vol. 1044, pp. 43–50, 2004.

[4] A. Fernández-García, E. Zarza, L. Valenzuela, and M. Pérez, "Parabolic-trough solar collectors and their applications," *Renew. Sustain. Energy Rev.*, vol. 14, no. 7, pp. 1695–1721, 2010.

[5] K. M. Knight, S. A. Klein, and J. A. Duffie, "A methodology for the synthesis of hourly weather data," *Sol. Energy*, vol. 46, no. 2, pp. 109–120, 1991.

[6] J. García-Barberena *et al.*, "Analysis of the influence of operational strategies in plant performance using SimulCET, simulation software for parabolic trough power plants," *Sol. Energy*, vol. 86, no. 1, pp. 53–63, 2012.

[7] J. Bonilla, L. J. Yebra, S. Dormido, and E. Zarza, "Parabolic-trough solar thermal power plant simulation scheme, multi-objective genetic algorithm calibration and validation," *Sol. Energy*, vol. 86, no. 1, pp. 531–540, 2012.

[8] X. Zhang *et al.*, "Solar Thermal System Evaluation in China," *Int. J. Photoenergy*, vol. 2015, 2015.

536005 Stochastic Optimization to Reduce Cost of Energy for Parabolic Trough Solar Power Plant for Different Weather Conditions

Abstract:

The need for clean and cheap renewable energy is on the rise. Solar energy is one of the cleanest and readily available technologies with almost zero carbon emissions. Optimizing the resources to produce efficient power at low costs is the need of the day. However, solar energy power plants face some uncertainties like the weather. Since the technologies are new, cost uncertainties are common. In this work, we optimize the cost of the solar thermal power plant of four different locations of USA (North East, North West, South East, South West) using the novel Better Optimization of Nonlinear Uncertain Systems (BONUS) algorithm. We use the System Advisory Model (SAM) system from NREL to model performance and economics of the power plant. Since this is a black box model, optimization and optimization under uncertainty becomes difficult. Unlike the deterministic optimization problem, in stochastic programming (or stochastic optimization), one must consider the probabilistic functional of the objective function and constraints. The generalized treatment of such problems is to use computationally intensive probabilistic or stochastic models instead of the deterministic model, inside the optimization loop. BONUS can circumvent the problems associated with black box models, computational intensity of sampling, and perturbation derivative costs. Instead of the stochastic modeling loop, BONUS uses a statistical reweighting approach to obtain the probabilistic information. The results of our work show that using BONUS algorithm; we get 41% - 47% of savings on the Levelized Cost of Electricity (LCOE) for Parabolic Trough Solar Power Plant. The four regional cities show different optimal designs, and the optimal Levelized cost of electricity (LCOE) varies significantly. For example, the optimal LCOE of New York is 57.42% more, Jacksonville is 38.52% more, and San Diego 17.57% more than that of Las Vegas.

528976 Stability of Platinum in the Electrochemical Environment: Reconstruction, Roughening, and the Third Peak

Abstract:

Platinum is a widely used catalyst in aqueous and electrochemical environments. It is used in fuel cells and electrolyzers, for example, to catalyze hydrogen oxidation, hydrogen evolution, and the oxygen reduction reactions [1,2]. The activity and selectivity of a catalyst is dictated by its surface structure, which can change in the reaction environment. In our prior work, we have used density functional theory to examine the surface energy (thermodynamic stability) of the low index Pt(111), Pt(100), and Pt(110) surfaces in the presence of adsorbed hydrogen, hydroxide, water, oxygen, surface oxide, and alkali metal cations, spanning the complete electrochemical window of water [3]. We found that the relative stability of these surfaces was altered in the electrochemical environment. While Pt(111) is the most stable surface in vacuum, the strong binding of hydrogen to Pt(100) and of hydroxide and water to Pt(110) make these surfaces more stable at low potentials and high potentials, respectively. We used these results to better understand the reconstruction of platinum electrodes. For example, cycling the potential of a Pt(111) electrode above 1.3 V_{RHE} results in the growth of islands expressing, primarily, (110) step sites [4, 5]; the Pt(110) surface is most stable at potentials where the surface oxide formed at high potentials (> 1.3 V_{RHE}) is reduced [3]. Similarly, the formation of (100) terrace sites upon square wave cycling of spherical nanoparticles to low potentials (below 0 V_{RHE}) [6] may be driven by the greater stability of the Pt(100) surface at low potentials due to its strong binding of hydrogen [3]. In this talk we will show how we have recently extended this work to include the surface energy of the Pt(553) and Pt(533) stepped surfaces. We have additionally considered the thermodynamics of subsurface hydrogen formation and hydrogen induced step-edge roughening. Using density functional theory, we find this step-edge roughening to be both thermodynamically and kinetically favorable. We compare closely to experiment, where it may be responsible for the appearance of the anomalous anodic "third peak" which can be seen in cyclic voltammograms on cycling a stepped or polycrystalline platinum electrode to low potentials (below -0.2 V_{RHE}) [7, 8].

1) H. A. Gasteiger, J. E. Panels, and S. G. Yan, "Dependence of PEM fuel cell performance on catalyst loading," ***Journal of Power Sources,*** vol. 127, pp. 162-171, 2004.

2) H. A. Gasteiger, S. S. Kocha, B. Sompalli, and F. T. Wagner, "Activity benchmarks and requirements for Pt, Pt-alloy, and non-Pt oxygen reduction catalysts for PEMFCs," ***Applied Catalysis B: Environmental,*** vol. 56, pp. 9-35, 2005.

3) I. T. McCrum, M. A. Hickner, and M. J. Janik, "First-principles calculation of Pt surface energies in an electrochemical environment: thermodynamic driving forces for surface faceting and nanoparticle reconstruction," ***Langmuir***, vol 33, pp. 7043-7052, 2017.

4) M. Wakisaka, S. Asizawa, H. Uchida, M. Watanabe. "In Situ STM Observation of Morphological Changes of the Pt(111) Electrode Surface During Potential Cycling in 10 Mm HF Solution," ***Phys. Chem. Chem. Phys,*** vol. 12, pp. 4184– 4190, 2010.

5) L. Jacobse, Y.-F. Huang, M. T.M. Koper, M. J. Rost, "Correlation of surface site formation to nanoisland growth in the electrochemical roughening of Pt(111)", ***Nature Materials***, vol. 17, pp. 277-282, 2018.

6) N. Tian, Z.-Y. Zhou, S.-G. Sun, Y. Ding, Z. L. Wang, "Synthesis of Tetrahexahedral Platinum Nanocrystals with High-Index Facets and High Electro-Oxidation Activity" ***Science***, vol. 316, pp. 732– 735, 2007.

7) K. Kinoshita, J. Lundquist, P. Stonehart, "Hydrogen adsorption on high surface area platinum crystallites", ***Journal of Catalysis***, vol. 31, pp. 325-334, 1973.

8) O. Diaz-Morales, T. J.P. Hersbach, C. Badan, A. Garcia, M. T.M. Koper, "Hydrogen adsorption on nano-structured platinum", ***Faraday Discussions***, Accepted Manuscript, 2018, DOI: 10.1039/C8FD00062J.

537022 Sustainable Synthesis of Electrocatalytic Bismuth-Based Core-Shell and De-Alloyed Nanoparticles

Abstract:

The dissolution of base metal cores made of Ni, Fe, and Co in the hot, acidic environment of proton exchange membrane fuel cells (PEMFCs) has impeded the use of novel core-shell nanoparticles (NPs) where the precious metals Pt and Ru are confined to surface. Since electrocatalysis occurs on surfaces, a substantial reduction in precious metal use could be achieved with core-shell NPs relative to pure systems, as long as the core remains stable during use. Bi is an abundant earth metal with a low environmental and human toxicity and water stability across a wide pH range, particularly under the acidic conditions of a fuel cell environment. These characteristics make Bi an ideal candidate for use as a durable and inexpensive core material to minimize precious metal use in highly dispersed core-shell Bi-Pt nanoparticle (NP) electrocatalyst in PEMFCs. Bi has been found to substantially promote direct alcohol fuel cell (DAFC) efficiency relative to pure Pd by mitigating catalyst poisoning due to irreversibly bound reaction intermediates such as CO and acetaldehyde. Unfortunately, the formation of Bi NPs is difficult and currently based on unsustainable methods employing toxic organic solvents, expensive organic precursors, and dangerous reducing agents such as hydrazine and lithium that frequently yield highly non-uniform structures. In this work, we describe an inorganic ligand metal redox synthesis approach which is used to synthesize a variety of well-defined core-shell as well as Pt-Bi alloy nanoparticle structures under sustainable, all-aqueous conditions. The approach uses $SnCl_3^-$ as both a reducing agent and inorganic stabilizing ligand to provide unprecedented control over the size and structure of these nanoparticles. Furthermore, the ligand itself exists in a variety of structures that can be used to control electrostatic assembly onto various substrates. The materials to be discussed in this presentation have never been previously reported and their synthesis is unexpected given both that pure Bi has well-known instability in high salt, low pH environment and that Bi-Sn ligand formation has not been previously recognized. The implications of such metal NP synthesis are transformative with respect to the practical realization and wide application of alternative energy technologies given its simplicity and sustainability. We will provide data showing that such structures can double the electrocatalytic oxygen reduction reaction mass activity and increase the specific activity by nearly an order of magnitude relative to pure systems (see graphic below). As important, we observe simultaneous mitigation of carbon monoxide poisoning due to the persistence of Bi and Sn on their surface. These observations are discussed relative to the surface distribution of active sites as characterized utilizing electrochemical adatom stripping.

536531 Atomically Dispersed and Nitrogen Coordinated Metal Site Catalysts for Oxygen Reduction in Acids

Abstract:

One of the bottlenecks hindering the wide implementation of polymer electrolyte fuel cells (PEMFCs) is the high cost and massive use of platinum (Pt) catalysts for boosting oxygen reduction reaction (ORR) kinetics at the cathode in PEMFCs. Fe-N-C catalysts, derived from pyrolysis of nitrogen, iron and carbon precursors together, have been the most promising candidates to replace Pt because of the abundance and low cost of such elements as well as the highly intrinsic ORR activity of Fe-N-C sites forming during the pyrolysis[1]. However, the major challenge for such Fe-N-C catalysts is to obtain sufficient Fe-N-C active sites catalysts to achieve high ORR activity, since undesired metallic species (i.e. Fe/Fe_3C) are easily produced during high-temperature pyrolysis due to the improper design of synthesis and the poor chemistry control in the preparation of catalysts, lowering the actual Fe-N-C active sites density of catalysts. To address this issue, we have developed an facile approach to prepare Fe-N-C catalysts with exclusively atomically dispersed Fe in Fe-N-C single-sites instead of containing metallic Fe particles by using well-defined Fe-doped metal-organic framework (MOF) precursors. The morphology of MOF precursors can be directly transferred into Fe-N-C catalysts with the retained shape after pyrolysis. Such controlled synthesis of precursors and resulting homogenous catalysts allow us to precisely control and tune the composition and morphology of Fe-N-C catalysts to understand how the structure and composition change of catalysts impact their ORR activity.

This presentation will discuss the critical roles of Fe content in the preparation of catalyst and particle size of catalysts on their ORR activity. The Fe content are controlled at the stage of Fe doped MOF precursor synthesis. Trace amounts of Fe was found to significantly boost ORR activity of catalysts, attributed by the highly-dispersed exclusive Fe-N-C active sites. The ORR activity of such Fe-N-C catalysts enhanced when increasing the density of Fe-N-C active sites at low range of Fe content. On the other hand, decreasing in ORR activity was observed when using high Fe content in the synthesis because of the formation of Fe particles in catalysts at high Fe content. Moreover, it is known that the particle size at nanoscale level critically governs the activity of catalysts. Taking advantages of this facile synthesis from MOF precursors, the Fe-N-C catalysts with size from 20 nm to 1000 nm are able to be prepared by adjusting the size of MOF precursors[2]. As a result, reduction of particle size in Fe-N-C catalysts allows us to increase the utilization of Fe-N-C active sites for ORR. Among studied catalysts, the catalyst with particle size around 50 nm shows the best ORR activity with a half-wave potential of 0.85 V *vs.* RHE, only leaving 30 mV gap with Pt/C (60 $\mu g_{Pt}/cm^2$) in 0.5 M H_2SO_4 along with the excellent stability. When the particle size in catalyst was further reduced to 20 nm, massive agglomeration of particles were found in the catalyst with a decrease in ORR activity. These high-performance Fe-N-C catalysts present a promising potential to replace Pt for ORR in future PEMFCs. In addition to extensively studied Fe-N-C catalysts, atomically dispersed and nitrogen coordinated Co and Mn sites catalysts will be briefly introduced in terms of their synthesis, characterization, electrochemical activity and stability in RDE, and fuel cell performance.

526772 Oxygen Reduction Reaction over a Novel 3D Pt-Supported Vertically Aligned Carbon Nanofiber

Abstract:

For both proton exchange membrane and direct methanol fuel cells, oxygen reduction reaction (ORR) is a critical, performance-determining step. While Pt has been established as a standard bearer in the field of ORR catalyst development, practical long-term applications require continuing effort to pursue active, stable, as well as affordable materials as alternatives to precious metals.

In this talk, carbon-based vertically aligned carbon nanofibers (VACNFs), essentially a stack of conical graphitic structures, will be discussed as highly promising ORR catalysts. In alkaline medium, at -0.4 V applied potential, the current density measured on VACNF cathode is twice more than that measured with commercial Pt catalyst (Pt/C). Molecular insights have been gained through the use of Density functional theory (DFT) calculations to: (1) reveal interactions of ORR intermediate species at the catalytically active sites; and (2) elucidate the origin of catalytic activities of VACNF under reaction conditions. Specifically, semi-periodic fishbone-like stacked graphene sheets, bare or terminated with H and OH, have been employed to understand the impact on ORR mechanism and corresponding reactivities. Bindings of O_2, O, and OH at various configurations obtained from DFT are then used as descriptors to assess catalyst performance against simple Pt (111) and Pt (211) surfaces.

In addition, Pt nanocatalysts supported on VACNF (Pt/VACNF), which can push the ORR current density even higher, will be discussed as well. Again, with DFT, a model with Pt anchored at the VACNF edges was proposed to investigate the functionalities of Pt/VACNF for alkaline ORR reactions. In this talk, both associative and dissociative ORR pathways will be analyzed based on the established models. In particular, the overall computed ORR thermodynamics are shown to be favored toward Pt/VACNF than Pt (111), consistent with experimental results. Nevertheless, the unique interactions of ORR intermediates such as O*, OOH*, and OH* with catalytic active VACNF edge sites, revealed by DFT calculations, suggest intriguing but also complex electrochemistries.

528877 PdCu Alloy Nanoparticles As Highly Active Electrocatalysts for Hydrogen Oxidation in Alkaline Electrolyte

Abstract:

Alkaline membrane fuel cells hold promise to reduce cost by employing a non-precious metal cathode catalyst, however, exploring more efficient anode catalyst to improving hydrogen oxidation reaction, which becomes sluggish in alkaline electrolyte, remains a great challenge. Herein, we report that the bcc-phase PdCu alloy nanoparticles synthesized via a wet-chemical route followed by a thermal-treatment exhibit surprisingly higher HOR mass and specific activities than those of Pd/C and Pt/C in a liquid alkaline electrolyte. HRTEM, XPS, HAADF-STEM element mapping and EELS scanning results show that the particle size, surface composition, and chemical state of PdCu nanoparticles do not change during applied thermal treatment. In-situ HE-XRD characterization reveals that low annealing temperature favors formation of a PdCu fcc structure, but at higher annealing temperatures, the bcc structure dominates PdCu NPs. HOR activity of PdCu catalyst can be attributed to bcc phase dominance. Density functional theory (DFT) computations further unravel the enhanced HOR activity on the bcc structure PdCu NPs originates from contributions from both H and OH adsorption. With the incorporation of Cu, the PdCu catalyst exhibits much weaker hydrogen adsorption strength than that on Pd surface. The PdCu fcc and bcc surfaces possess very similar H binding strength, however, the PdCu bcc surface features much stronger OH binding as compared to that on PdCu fcc surface, which is closer to that on the most HOR active Pt model surfaces. Therefore, the PdCu-bcc phase exhibits much higher HOR activity than its fcc counterpart due to synergistic optimization of both H and OH binding strength.

524239 Recent Developments in Electrochemical Synthesis of Hydrogen Peroxide

Abstract:

Recent Developments in Electrochemical Synthesis of Hydrogen Peroxide

Samira Siahrostami

samiras@stanford.edu

[1] SUNCAT Center for Interface Science and Catalysis, Department of Chemical Engineering, Stanford University, 443 Via Ortega, Stanford, California 94305, United States

Hydrogen peroxide (H_2O_2), is an environmentally benign and powerful chemical oxidant being able to eliminate a wide range of harmful bacteria and inorganic pollutants in water. Electrochemical production of hydrogen peroxide (H_2O_2) offers a sustainable way for local synthesis and onsite applications.[1–3] In this talk, I summarize the recent developments and efforts in electrochemical synthesis of hydrogen peroxide guided by both theory and experiment. [1–6]

References

(1) Siahrostami, S.; Verdaguer-Casadevall, A.; Karamad, M.; Deiana, D.; Malacrida, P.; Wickman, B.; Escudero-Escribano, M.; Paoli, E. a; Frydendal, R.; Hansen, T. W.; Chorkendorff, I.; Stephens, I. E. L. S.; Stephens, I. E.; Rossmeisl, J. *Nat. Mater.* **2013**, *12* (12), 1137–1143.

(2) Shi, X.; Siahrostami, S.; Li, G.; Zhang, Y.; Chakthranont, P.; Studt, F.; Jaramillo, T.F.; Zheng, X.; Nørskov, J. K. *Nat. Commun.* **2017**.

(3) Lu, Z.; Chen, G.; Siahrostami, S.; Chen, Z.; Liu, K.; Xie, J.; Liao, L.; Wu, T.; Lin, D.; Liu, Y.; Jaramillo, T. F.; Nørskov, J. K.; Cui, Y. *Nat. Catal.* **2018**.

(4) Chen, S.; Chen, Z.; Siahrostami, S.; Kim, T. R.; Nordlund, D.; Sokaras, D.; Nowak, S.; To, J. W. F.; Higgins, D.; Sinclair, R.; Nørskov, J. K.; Jaramillo, T. F.; Bao, Z. .

(5) Chen, Z.; Chen, S.; Siahrostami, S.; Chakthranont, P.; Hahn, C.; Nordlund, D.; Dimosthenis, S.; Nørskov, J. K.; Bao, Z.; Jaramillo, T. F.; Wilcox, J.; Bao, Z. N.; Rossmeisl, J. *React. Chem. Eng.* **2017**, *1*, 53.

(6) Siahrostami, S.; Li, G.-L.; Viswanathan, V.; Nørskov, J. K. *J. Phys. Chem. Lett.* **2017**, 1157–1160.

542620 The Impact of Shale Gas and Oil on the Chemical Industry

Abstract:

Lower alkanes and aromatics are very useful as both fuels and chemical raw materials. Occasionally supply disruptions constrained availability but also encouraged investigation of alternative feedstocks and chemistries. Recent new technologies for the production of unconventional gas and oil sources from tight shale formations significantly increased availability of both methane and condensate, radically altered process economics, resulted in the construction of much new intermediates capacity, and multiplied the ways traditional and alternative chemistries might be exploited. This presentation will outline historically how natural gas components became useful as chemical feedstocks, how government regulations occasionally led to perturbations in availability and price, how supply crises led to the development of alternative chemistries, catalysts, and ultimately the exploitation of shale resources. More specifically, will be discussed how excess coproducts, especially condensate, from increasing shale gas and oil production for energy are having significant unexpected impacts on the chemical process industries and opportunities for new process chemistry research.

541186 Sustainable Energy and Chemicals: Past, Present, and Future

Abstract:

The last 40 years has seen the U.S. energy situation vacillate from gasoline rationing and concerns over Peak Oil, to projected energy and feedstock prices that were noncompetitive in global markets for chemicals production, to the economic revolution of frack gas and light tight oil. The importance of low cost energy and feedstocks in driving economic opportunity will be discussed relative to this history, along with the importance of sustainability in decision making to gain stakeholder buy-in and acceptance. Strategies for managing career paths and competencies despite changing targets and high uncertainty will be addressed.

542618 Disruptions: What the Future May Hold

Abstract:

Disruptive forces are all around us. They change the way we communicate and work. They change the way we develop goods and services. They also change the types of products that are manufactured. Examples such as the way in which we produce and consume energy or the feedstocks we use to produce product have seen many changes over the course of the last few decades, and these disruptions are expected to continue. The chemical engineering discipline and the expectations for a chemical engineerâ€™s skills has also changed and adapted as a result of these disruptive forces. A discussion of the broad themes associated with these forces and the implications that they may have to chemical engineers in the future will be discussed.

542621 Geopolitical Factors Influencing the Evolution of the Chemical Industry

Abstract:

The basic chemicals and plastics industry is built on a simple principle: find a cheap feedstock and convert it into something more valuable. However, the right feedstock depends on place and time. Certain policies and actions of various governments have influenced the evolution and profitability of the chemical industry during the past 50 years, sometimes in surprising ways. For example, high oil prices and US government funding encouraged the development and commercialization of coal gasification technology during the late 1970s and early 1980s. This enabled China to develop a competitive basic chemical industry using coal as a feedstock during the last 25 years. Similarly, an abrupt increase in oil prices during the first decade of the 21st century encouraged the present shale gas revolution, causing a major shift in the economy of producing basic chemicals and polymers in the US. If history is our guide, then what governments choose to fund and regulate may be as important as market forces in determining the shape of our industry and profession in the future.

542622 Agility & Resilience: How to Maintain Career Competitiveness in the Changing Chemical Industry

Abstract:

The world around is changing. The shift to protectionism, global instability and price volatility of raw materials and feedstocks has the potential to wreak havoc on the chemical industry. With the revolution of shale gas in North America and the sudden drop in feedstock costs, the chemical sector in the US has reasserted its historic dominance. Learn how the role of a chemical engineer can contribute to industry growth and resilience, while adapting to allow for personal agility throughout the career life cycle.

525237 Metal-Modified Transition Metal Nitride Electrocatalysts for Oer, HER, and Other Reactions

Abstract:

Transition metal nitrides are attractive electrocatalytic support materials for similar reasons as their extensively studied carbide counterparts â€“ they are particularly useful for reducing precious metal loading, compared to a carbon support. Nitrides, however, have the distinct advantage of avoiding the carbonaceous overlayer resulting from typical carbide synthesis techniques. The bare nitride surface allows ad-metals to directly contact nitrides, which can improve performance and strengthen correlations with density functional theory calculations on model surfaces. This work utilizes these advantages to develop low-cost and effective electrocatalysts for various fuel cell and electrolysis reactions. Specifically, (1) iridium-metal nitride core-shell particles are used to reduce Ir loading for acidic OER by 50 percent, (2) monolayer platinum on niobium nitride and tungsten nitride thin films are demonstrated as effective HER catalysts in alkaline and acidic electrolytes, and (3) systematic stability studies are conducted on nitride thin films to evaluate their utility for various reactions across wide potential-pH regimes.

525521 Combining Electrochemistry and Surface Science to Identify Electrocatalytic Structure-Property Relationships

Abstract:

The oxygen evolution reaction (OER) is the anodic process associated with the electrochemical CO_2 reduction reaction (CO_2RR). State-of-the-art OER electrocatalysts contain precious metals, such as Iridium, which introduce substantial cost-barriers to large-scale CO_2RR deployment. Current efforts have focused on identifying inexpensive, earth-abundant transition metals with satisfactory OER activity. Nickel, cobalt, iron and bimetallic CoFe and NiFe electrocatalysts have shown remarkable OER activity in basic conditions, but atomic-level understanding of how catalyst shape, size, and composition are still lacking in the literature.

Here, we present an approach that couples surface science and electrochemistry to synthesize, characterize and evaluate well-defined transition metal electrocatalysts. We demonstrate this technique with well-defined Fe_2O_3 nanocatalysts that show outstanding, structure-dependent mass activity with values exceeding 10^5 A/g nanocatalysts were grown on Au(111) supports in ultra-high vacuum (UHV) and characterized with scanning tunneling microscopy (STM) and x-ray photoelectron spectroscopy (XPS). Our synthetic approach provided extremely precise structural control and we varied catalyst structures from near monolayer coverage to discrete, isolated nanoparticles. High-resolution STM imaging quantified the nanocatalyst morphology, including size, thickness, total surface area, and perimeter-to-surface-area ratios. Samples were translated from UHV and tested for electrochemical OER in 0.1M KOH solutions. Post-reaction STM and XPS characterization revealed structural stability during OER and we identified a relationship between particle size, density of perimeter Fe-oxide(hydroxide) sites, and OER activity. This represents some of the clearest identification of reactive site location in any electrocatalyst system to date, and the results confirm long-held hypothesis that undercoordinated or â€œedgeâ€ sites are responsible for catalytic activity. Ongoing computational efforts within Density Functional Theory (DFT) will provide atomic-level details about the reaction energetics and mechanisms at various active sites to explain experimentally observed activity trends. Future efforts will be directed towards more complex OER systems to probe how catalyst structure, composition, and active site location impact electrocatalytic activity.

534739 Active Structures and Species of Modified Oxide Catalysts for the Oxygen Evolution Reaction (OER)

Abstract:

A major cause of energy losses in photocatalytic systems for solar fuels, and also in a number of other emergent technologies such as rechargeable metal-air batteries, water splitting devices, and unitized regenerative fuel cells, is the oxygen evolution reaction (OER). The low energy efficiencies of OER are the result of its sluggish reaction kinetics and large overpotentials. Significant improvements to the OER activity of transition metal oxides (TMOs) have been made by tailoring the morphology and crystal structure of the catalysts, incorporating dopants, as well as using conductive supports. However, clear structure-activity correlations remain elusive because of the complex composition and structure of TMO catalysts. Further understanding of these relationships for these promising catalysts for oxygen evolution may lead to additional developments that will enable TMOs to replace precious metal-based OER catalysts (e.g. IrO_x and RuO_x).

Insights from two recent examples of our studies will be discussed. We have utilized a range of spectroscopic techniques for characterization of Ce-modified copper oxide (CuO_x) [1] and Ni-modified cobalt (oxy)hydroxides [2] to reveal the OER active species and structures of these catalysts. In the case of Ce-modified CuO_x, Ce incorporation (6.9 at%) into CuO_x led to 3.3 times greater OER activity compared to pure CuO_x and this was coincident with significant structural changes due to an increasing amount of disorder. By combining X-ray photoelectron and Raman spectroscopy techniques, a strong correlation between OER performance with tetravalent Ce (Ce^{4+}) ions was observed up to a concentration corresponding to CeO_2 phase segregation. We propose a strong promoting effect of Ce^{4+} for OER in this system. In the case of Ni-modified CoO_xH_y, operando Raman spectroscopy was used to reveal a drastic transformation of a spinel Co_3O_4-like structure into a more active (oxy)hydroxide structure under applied potential. Such a transformation was only observed in the presence of uniformly distributed Ni ions. These two examples, i.e. the promoting effect of Ce^{4+} and the formation of active OER structures in Ni-modified CoO_xH_y, reveal the importance of chemical state and local structure considerations for the rational design of improved oxide-based OER catalysts.

[1] Z. Chen, C.X. Kronawitter, X. Yang, Y.-W. Yeh, N. Yao, and B.E. Koel, *Phys. Chem. Chem. Phys.*, **2017**, *19*, 31545-31552.

[2] Z. Chen, L. Cai, X. Yang, C. Kronawitter, L. Guo, S. Shen, B. E. Koel, *ACS Catal.*, **2018**, *8*, 1238â^'1247.

537033 Lowering the Charge Overpotentials in Li-O_2 Battery By Tailoring the Oxygen Reduction and Evolution Reaction Energetics Using Non-Precious Metal Oxide Electrocatalysts

Abstract:

Lithium-oxygen (Li-O_2) batteries are considered as the next generation energy storage technology with vast potential to contribute to electrification of the transportation sector. Although these systems are promising, several factors have resulted in performance far below the theoretical potential, including sluggish oxygen evolution reaction (OER) kinetics leading to high overpotential losses during charging.[1] We report a decrease in these overpotential losses achieved by perturbing the energetics associated with oxygen reduction reaction (ORR) and OER in aprotic Li-O_2 cathodes containing non-stoichiometric mixed metal oxide electrocatalysts with mixed electronic and ionic conducting properties.[2-3] A combination of well controlled synthesis, electrochemical studies and detailed characterization by Raman spectroscopy, titration, and differential electrochemical mass spectrometry show that these non-stoichiometric mixed metal oxide nanostructures change the energetics associated with ORR as compared to carbon-only electrodes by stabilizing Li-deficient intermediate species (> 50% more LiO_2 is detected in the presence of the catalyst), which consequently leads to lower OER overpotentials during charge. In addition, the presence of the catalyst also reduces the charge overpotential losses by lowering the potential required to dissociate Li_2O_2 during charge. These insights are key in providing new avenues for improving the efficiency of Li-O_2 batteries via tailoring the ORR/OER energetics using well-defined nanostructured, non-precious metal oxide electrocatalysts.

References:

(1) Lu, J.; Li, L.; Park, J. B.; Sun, Y. K.; Wu, F.; Amine, K., Aprotic and Aqueous Li-O_2 Batteries. *Chem. Rev.* **2014,** *114*, 5611-5640.

(2) Gu, X.-K.; Samira, S.; Nikolla, E., Oxygen Sponges for Electrocatalysis: Oxygen Reduction/Evolution on Non-stoichiometric, Mixed Metal Oxides. *Chem. Mater.* **2018,** DOI:10.1021/acs.chemmater.8b00694.

(3) Nacy, A.; Ma, X. F.; Nikolla, E., Nanostructured Nickelate Oxides as Efficient and Stable Cathode Electrocatalysts for Li-O-2 Batteries. *Top. Catal.* **2015,** *58*, 513-521.

538815 Density Functional Theory Study of Oxygen Evolution Reaction on Specific Terminated Facets of Perovskite Oxides

Abstract:

Developing efficient oxygen evolution reaction (OER) electrocatalysts is vital to the new generation of electrochemical storage and conversion devices such as electrolyzer and metal-air batteries. In order to address the longstanding problems with high cost and poor stability of noble metal electrocatalyst, this study selects perovskite oxides (ABO_3) as a model system because of their low-cost and high versatility in compositions and tunable electronic structures. In this talk, we will present the fundamental mechanism whereby the electrocatalytic activity on various facets of perovskite oxides in oxygen evolution reaction using the ab initio density functional theory calculations. Elemental doping on A-site and therefore, the fundamental role and effect of doping on the performance of OER reactions on the various perovskite surfaces will also be discussed. Free energy of electrochemical reactions will also be discussed by referencing to the computational hydrogen electrode model. Some of the results will also be compared with the experimental findings.

539677 Core-Shell Nanoparticles for Efficient Oxygen Evolution Electrocatalysis in Alkaline and Acidic Media

Abstract:

By storing excess renewable energy in chemical bonds, water electrolyzers provide a promising means of renewable energy storage and clean hydrogen fuel generation toward a sustainable energy infrastructure. Currently, the slow kinetics of the oxygen evolution reaction (OER) occurring at the electrolyzer anode limit device efficiencies, necessitating the development of high performance electrocatalysts. Core-shell and surface-leached nanostructures provide promising strategies to achieve enhanced OER catalysis through electronic and geometric structure alterations. This presentation will explore these strategies in alkaline and acidic electrolyte through the discussion of two model systems: Au@M_xO_y (where M=Ni, Co, Fe, and CoFe) core@shell nanoparticles for alkaline OER and surface-leached crystalline strontium iridate nanostructures for acidic OER. In alkaline media, the beneficial effects of Au-supported catalysts were translated to core-shell Au@M_xO_y nanoparticles, demonstrating a universal activity enhancement with the inclusion of the Au-core. The highest activity particles, Au@$CoFeO_x$, demonstrated an overpotential of 328 ± 3 mV over a 2 h stability test at 10 mA cm^{-2}, illustrating that strategically combining Au support and mixed metal oxide effects in a core-shell nanoparticle architecture is a promising avenue for practical catalyst design. In acidic media, the effect of crystal structure and stoichiometry on the activity and stability of crystalline strontium iridate (e.g. Sr_2IrO_4, $SrIrO_3$, and Sr_4IrO_6) nanostructures is evaluated in comparison to commercial Ir/C nanoparticles. In-situ leaching results in an IrO_x-rich surface on a strontium iridate core. Detailed microscopy and spectroscopic evaluation before and after electrochemistry is provided to elucidate performance trends and highlight structure-activity-stability relationships. These two systems demonstrate the potential catalytic benefits of core-shell motifs resulting from both direct chemical synthesis and post-synthetic *in-situ* modification towards the OER in alkaline and acidic electrolyte.

543554 Gas Capture by Ionic liquids

Abstract:

With the rapid development of the global industrialization process, a large number of gases, such as CO_2, SO_2, NH_3 and H_2S, have been produced. Among of them, CO_2 and SO_2 have caused air pollution problems, resulting in a serious harm to the environment and human health. On the other hand, CO_2 and SO_2 are also important source for synthesizing chemicals. Recently, ionic liquids (ILs) have been regarded as one of the most promising alternative solvents for capturing CO_2 or SO_2 due to their special physicochemical properties, such as negligible vapor pressure and designable character. For example, it was found that N-butylpyridinium thiocyanate ([C_4Py][SCN]) can absorb 0.841 gSO_2/g IL at 293 K and 0.10 MPa[1].

In this work, the mechanism of CO_2/SO_2 absorption in different kinds of ILs were systemically studied by molecular simulations. The microscopic structure, interaction energy and dynamic properties of the mixtures of ILs and CO_2/SO_2 were investigated. The results of interaction energy and fractional free volume of ILs showed that unlike non-polar gas CO_2, the interaction energy of cation-anion and free volume do not play a dominant role in SO_2 absorption. The solvation of SO_2 in ILs is controlled by enthalpy. Both the cations and anions play an important role in SO_2 solubility. Based on the above studies, a serial of ILs were synthesized for capturing CO_2/SO_2. Moreover, the energy and economic analysis of whole process of CO_2 capture with ILs were simulated. It was found that the IL process saved about 15% regeneration heat energy consumption and 11.5% total capture cost compared to the MEA process.

Acknowledgment

This work was supported by the National Natural Scientific Fund of China 51674234.

Reference

1. J. Zeng, H. S. Gao, X. C. Zhang, H. F. Dong, X. P. Zhang, S. J. Zhang, *Chem. Eng. J*, 2014, 251, 248-256.

533203 Thermoeconomic Optimization of Reverse Brayton Cycle Based Cryocooler for HTS Power Transmission Cable

Abstract:

In the era of the global sustainable energy crisis, an efficient power transmission system is required to meet the increasing power demand. After the discovery of superconductivity in 1911, many materials with zero resistance at a temperature below 25 K, known as Low-Temperature Superconductor (LTS), were discovered and a possible alternative for efficient power transmission. However, the cost of these materials and cooling system to maintain them in the superconducting state was very high. Thus, it is practically not feasible to use such materials for the transmission system. In 1986, the discovery of High Temperature Superconductors (HTS), which can be maintained in the superconducting state using liquid nitrogen, made the practical and economically feasible for use HTS power cable. These cables have to be maintained in a temperature range of 65 to 80 K with a cooling load of 2 to 5 kW for 1 km long transmission cable. Reverse Brayton cycle based Cryocooler (RBC) is the most suitable cooling system because of a wide range of cooling capacity at different temperature level. However, the capital and operational costs of RBC is high compared to the cryogenic roadmap (Gouge, 2004). The current cost of refrigeration for HTS devices is five to ten times of cryogenic roadmap ($25 per watt of refrigeration at 65-80 K). Hence, a thermoeconomic optimisation of RBC has been carried out to enhance the performance of the system from the thermodynamic and economic point of view.

Fig. 1 Reverse Brayton Cycle based cryocooler for HTS power transmission cable.

In the present study, a reverse Brayton cryocooler consists of a compressor, main heat exchanger, turboexpander and load heat exchange as shown in Fig.1, has been investigated and optimised for 10 kW cooling load at 65 K. A commercial process simulator Aspen HYSYS V8.6 has been used for the study. Helium, more efficient than neon (Dhillon, 2017) has been used as working fluid for the cryocooler. The thermodynamic and economic models have been developed for analysing of the cryocooler. It is observed that the exergy efficiency of the system has maxima at 1.9 pressure ratio whereas cost per hour has manima at 2.3 pressure ratio as shown in Fig. 2. The inlet pressure and pressure ratio of the compressor have been varied to optimise the system process parameter. The effect of design parameters of equipment on the thermoeconomic performance of the system has been studied. The results obtained for RBC are presented in the present work and may be helpful for engineers and scientist to reduce the cost of RBC cryocooler without compromising the performance of the system.

Fig. 2 Variation of cost per hour and exergy efficiency of RBC with different pressure ratios

1. Gouge, M. Cryogenic cooling technology for HTS electric machinery, 2004. IEEE Power Engineering Society General Meeting, 1-3, DOI: 10.1109/PES.2004.1373239.

2. Dhillon, A. K., Dutta R. and Ghosh P. Exergetic analysis of reverse Brayton cryocooler for different cooling loads at 65 k for HTS cables, 2017. 14th Cryogenics 2017 IIR International Conference, DOI: 10.18462/iir.cryo.2017.027

530055 Technology Readiness Levels As a Framework for Techno-Economic Assessment

Abstract:

Techno-economic assessment (TEA) is an important basis for decisions about research, development and deployment (RD&D). Methods applied in TEA depend heavily on the availability of data which is closely linked to the development progress. Technology readiness analysis and especially its rating scheme, technology readiness levels (TRL), are popular concepts for rating the maturity of RD&D according to available data. TRLs can therefore serve as a framework that facilitates the choice of methods along RD&D by sorting them according to their required input data.

TRL rating is widely adopted in the evaluation of RD&D projects by government (*e.g.* policy makers) [1,2], industry and academia. However, as the TRL concept was developed for space exploration [3], a domain that integrates a variety of disciplines from mechanical, electrical and chemical engineering to aviation and medicine, TRL guidelines and criteria remain unspecific to technologies, which can lead to misinterpretation and overly optimistic evaluation. A variety of TRL definitions evolved over the years, for example for biomanufacturing [4] or for the defense sector [5], defining specific TRL criteria for their respective domains. The need for a specification of TRL analysis for the chemical industry was identified in a report of the European association of research and technology organizations (EARTO) [6].

In this contribution, we summarize existing TRL concepts and propose adaptions and extensions to criteria for TRL rating in the chemical industry: Firstly, TRLs are characterized by the description of the project phase, its tangible deliverables and other qualitative criteria. Secondly, for a more objective and comprehensible TRL, specific indicators representing known information of various aspects of chemical engineering such as reaction engineering or process development are proposed. In addition, quantitative indicators such as the production capacity are often used and therefore discussed for different types of processes in this contribution.

Furthermore, we outline how methods of (capital) cost estimation and profitability calculation can be sorted by TRL in order to give guidance for sound TEA.

We gratefully acknowledge the financial support by the European Institute of Innovation and Technology (EIT) and the German Federal Ministry of Education and Research (BMBF).

[1] European Commission. EN HORIZON 2020 WORK PROGRAMME 2016 â€“ 2017, General Annexes (European Commission Decision C (2017) 2468 of 24 April 2017), Annex G, Technology Readiness Levels (TRL); 2017.

[2] BMBF, Definition des Technologischen Reifegrades [â€¦] accessed on 08.02.2017 from https://www.ptj.de/lw_resource/datapool/_items/item_5502/definition_des_technologischen_reifegrades.pdf

[3] S. R. Sadin, F. Povinelli, R. Rosen, contribution at 39thIAF Congress, Bangalore, India, Oct. 8-15, 1988

[4] NSF Engineering Research Center for Biorenewable Chemicals (CBiRC). Technology Readiness Levels https://www.cbirc.iastate.edu/industry/technology-readiness-levels/ (accessed Oct 31, 2017).

[5] U.S. DEPARTMENT OF DEFENCE. Technology Readiness Levels in the Department of Defense (DoD) https://www.army.mil/e2/c/downloads/404585.pdf (accessed Oct 31, 2017)

[6] EARTO, The TRL Scale as a Research & Innovation Policy Tool, EARTO Recommendations, 2014

536017 Techno-Economic Modeling and Optimization of Catalytic Reactive Distillation for Bio-Oil Upgradation

Abstract:

The most significant benefit of pyrolysis bio-fuel is its availability from virtually every waste biological materials, which provides a double edge in sustainable waste management in addition to meeting our energy requirements. However, unfavorable characteristics of crude bio-oils received from pyrolysis process compared to petroleum fuels means that it must undergo upgradation before using in combustion engines. Reactive distillation (RD) is emerging as a powerful technology for these chemical reactive separation processes. RD, which has tremendous advantages over traditional reaction and separation unit operations including handling of unfavorable chemical and phase interactions, enhancing conversion and selectivity, circumventing azeotropic mixtures and reducing capital and energy costs, has been exploited for the reactive separation of various equilibrium controlled chemical reaction systems. Pyrolysis bio-oil upgradation is usually carried out by transesterification and esterification reactions with an alcohol and RD is suited pefectly to these types of equilibrium controlled reactions.

Here, equilibrium RD process simulation and economic evaluation was carried out for the esterification of n-butanol with complex mixtures of several fatty acids, phenolics and water (represent pyrolysis crude bio-oils) using Aspen PLUS process simulator and Aspen Process Economic Analyzer, respectively at both atmospheric and high pressure (10 bar). Complete design and optimization was carried out using evolutionary techniques from simplest system (ordinary distillation) to the most rigorous one (complex RD) with introduction of progressive complexities. RD simulation was performed using UNIQUAC as base-property method in Aspen RADFRAC module by minimizing Gibbs free energy. Prior to distillation design, esterification reaction kinetics and binary/ternary interactions were analyzed using Aspen RGibbs reactor module and Aspen Property PLUS, respectively. Simple distillation column was designed and optimized to obtain the initial trial parameters to be used in RD simulation, which was initially optimized only for acetic acid-water (75% acetic acid-25% water) mixture before RD design for complex fatty acid mixtures based on techno-economic feasibility and optimization. The reaction of phenol with fatty acids was incorporated and phenol was found to react with formic acid entirely, which increases the reaction conversion 5-10% due to its favorable effects. The conversions for the esterification reactions were found to be 88-99% for various simulated bio-oils with phenol composition 5-20% for butanol:acid (B:A) ratio of 2.5, and water was almost completely separated from the ester products which should increase the fuel quality. Higher water percentage in bio-oil feed was found to reduce acid conversion and ester separation. A reflux ratio of 0.95, distillate to feed ratio 0.505 and total 18 stages (3 rectifying, 9 reactive and 4 stripping) with n-butanol and bio-oil feed stage of 4 and 13, respectively, was found to be optimum at 1 bar based on reaction conversion, ester separation, heat duties and capital and operating costs. Effects of B:A feeding ratio and water percent in bio-oil is also analyzed and B:A = 2.25 was found to be optimum for 25% water in bio-oil. These works and optimized parameters can serve as a design platform for pyrolysis bio-oil upgradation to transportation fuels.

Reference:

1. Khan, M.A., Adewuyi, Y.G. High pressure reactive distillation simulation and optimization for the esterification of pyrolysis bio-oil. Process Engineering Journal, 1 (2017) 73-85.
2. Khan, M.A., Adewuyi, Y.G. Simulation of reactive distillation for esterification of pyrolysis bio-oil. The Proceeding of AIChE Annual Meeting, 2013, San Francisco, California, USA.

536658 Superstructure Optimization of Membrane-Based Carbon Capture Systems

Abstract:

This work presents the optimal design, configuration and operating conditions of a membrane-based post-combustion CO_2 capture process using computationally identified mixed matrix membranes. A new mixed integer nonlinear model is presented, which includes discrete and continuous decisions to minimize the cost of electricity (COE). The simultaneous optimization of process configuration, membrane materials, and operating conditions ensures rigorous exploration of the entire feasible design space.

Membrane systems often consist of multiple stages of membranes and compressors with intercooling. Flue gas source plays a very important role, since the CO_2 concentration highly affects the final plant configuration. While a simple multi-stage configuration may be optimal for concentrated CO
integrated with the boiler is often more effective for coal fired power plants.[1] Given the importance of finding the right configuration, two approaches have been used to optimize these systems. The first approach uses simulation-based optimization that leverages commercial process simulators such as Aspen Custom Modeler or gPROMS to develop rigorous models[2,3,4]. Previous work has used the Framework for Optimization, Quantification of Uncertainty and Surrogates (FOQUS)[2] to optimize the design and operating conditions of specific process configurations. In these approaches, the number of potential configurations that can be analyzed is limited since each must be set up individually.[3,4]

The second approach focuses on superstructure-based mathematical optimization of multi-stage membrane systems, which often requires simplified models for the equipment.[5,6] The advantage of the second approach is that the configuration as well as the design and operating conditions may be optimized simultaneously, resulting in more cost-effective plants.

Our work leverages efforts at NETL for developing new materials for membrane-based post-combustion capture processes.[2,3,4,7] An advanced superstructure-based process optimization model is developed using rigorous mathematical models for compressors, pumps, mixers, and heat exchangers and reduced order models for liquefier columns and property calculations. The current work makes use of results from a recent study[4], which analyzed a large number of hypothetical mixed matrix membrane materials for carbon capture. This work addresses the aforementioned limitations by analyzing multiple potential advanced process configurations and the selection of the membrane material from a wide set of potential options while optimizing the plant operation.

[1] Merkel T. C., Lin H., Wei X., Baker R. (2010). Power plant post-combustion carbon dioxide capture: An opportunity for membranes. Journal of Membrane Science. 359, 126-139.

[2] Miller D. C., Agarwal D., Bhattacharyya D., Boverhof J., Cheah Y. W., Chen Y., Eslick J., Leek J., Ma J., Mahapatra P., Ng B., Sahinidis N. V., Tong C., Zitney S. E. (2016). Innovative computational tools and models for the design, optimization and control of carbon capture processes, Computer Aided Chemical Engineering, 38, 2391-2396.

[3] Morinelly J. E. M., David C., 2012, Post-Combustion Gas Permeation Carbon Capture System Models, AIChE Annual Meeting, Pittsburgh, PA, October 28-November 2, 2012.

[4] Budhathoki S., Ajayi O., Steckel J.A., Wilmer C.E. (2018). High-throughput computational prediction of the cost of carbon capture using mixed matrix membranes. In preparation.

[5] Hasan M. M., Baliban R. C., Elia J. A., Floudas C. A. (2012). Modeling, Simulation, and Optimization of Postcombustion CO_2 Capture for Variable Feed Concentration and Flow Rate. 1. Chemical Absorption and Membrane Processes. Industrial and Engineering Chemistry Research. 51, 15642-15664.

[6] Arias A. M., Mussati M. C., Mores P. L., Scenna N. J. (2016). Optimization of multi-stage membrane systems for CO_2 capture from flue gas. International Journal of Greenhouse Gas Control. 53, 371-390.

[7] Zamarripa M. A., Ajayi O., Matuszewski M., Miller D. C., 2017, Superstructure-based optimization of membrane-based carbon capture systems. AIChE Annual Meeting, Minneapolis, MN, October 28-November 2, 2017.

543630 Making Hydrogen from Water with a Protein Organized Electrode: Ultra-High Utilization of Noble Metal in Proton Exchange Membrane Electrolysis for Capital Cost Reduction

Abstract:

Renewable hydrogen could play a critical role in low-carbon energy systems and in fulfilling intermittent energy needs. Proton exchange membrane (PEM) electrolysis technology for hydrogen production is safe, simple, and highly efficient. PEM electrolysis has ability to be integrated in a variety of energy systems and infrastructures, such as transportation, renewable electricity, and biogas upgrading. It also has the ability to address seasonal-variation in long timescales. But, one of the major challenges remains the high cost of noble-metal electrodes in the membrane electrode assembly (MEA). Conventional catalysts like platinum and iridium are required to facilitate the hydrogen evolution reaction (HER) and the oxygen evolution reaction (OER) at electrodes for superior stability in long-term. The contribution of catalyst cost in the MEA becomes a more significant fraction of the capital cost when the production is scaled up.

To reduce the capital cost, we have developed protein-integrated electrodes which are designed to allow lower amounts of noble metal catalyst while maintaining high electrolysis performance. The purpose of our designed protein, elastin-like polypeptides (ELP), is to organize ionomer networks and create efficient pathways between the ionomer and catalyst sites, thereby significantly improving catalyst utilization. A preliminary cost analysis indicates that even at current prices for engineered proteins and using reasonable ratios of protein to catalyst particles (~6.5 metal:protein by mass), a 10X reduction in noble metal could be achieved for the same cost. Considering this technology is expected to lower loadings by more than 10X, and that the future cost to produce these proteins will decrease while the cost of noble metals will increase, the technology has great economic potential. For example, if the cost of protein manufacturing drops by a reasonable 25%, and a 20X reduction in metal loading is achieved, the total electrode cost would be 60% lower. ELP is suitable as electrode organizing molecules for its flexibility in design, modular format, and manufacturability. Our results using a quartz crystal microbalance and a transmission electron microscope show a robust binding of ELP to platinum catalyst and ionomers which affects the resulting ionomer-catalyst structures. This system provides an innovative, multidisciplinary solution for improving catalyst utilization in membrane electrode assemblies.

522132 Design Optimization and Economic Evaluation of LNG/NGL Integrated Processes Under Lean Feed Composition

Abstract:

The economic evaluation of the integrated liquefaction and natural gas liquids (NGL) recovery process for the lean feed is required to meet recent increasing demand for lean gas reservoirs. This paper performs economic analysis for the conventional co-production of liquefied natural gas (LNG) and NGL process compared to the proposed simplified integrated process with consideration of the LNG higher heating value (HHV) specification under the lean feed condition. The dual mixed refrigerant (DMR) cycle was selected for liquefaction part due to high efficiency reported by previous studies. The schemes were built by Aspen HYSYS commercial software and the object function was optimized with a genetic algorithm (GA). The results show that the simplified integrated process has economic advantages compared to the conventional integrated scheme under the lean feed composition, which shows different results in terms of previous studies conducted by normal or rich feed condition. Additionally, the capital cost can be reduced remarkably by simplifying NGL recovery section with almost no loss of liquefaction efficiency in the proposed simplified integrated process.

516965 Electrochemical Deposition of Sr and Ba into Liquid Bi from Molten Salt Electrolytes

Abstract:

Electrochemical deposition of Sr and Ba into liquid Bi metal was investigated in LiCl-KCl-$SrCl_2$-$BaCl_2$ electrolytes at 500 °C as a means to separate stable alkaline-earth ions from the eutectic LiCl-KCl utilized for recycling used nuclear fuel, by leveraging the strong chemical interactions between alkaline-earth metals and liquid Bi. The liquid Bi electrodes were subjected to cathodic discharge in eutectic LiCl-KCl with the addition of 5 mol% total of $SrCl_2$ and/or $BaCl_2$. The use of Bi resulted in co-deposition of Sr (2.0–6.5 mol%), Ba (4.1–12.8 mol%), and Li (5.9–16.2 mol%). The observed co-deposition was supported by thermodynamic analyses of electrode potentials by incorporating the experimentally determined activity values of each alkali/alkaline-earth metal in Bi. Furthermore, the contents of $SrCl_2$ or $BaCl_2$ in the electrolyte was varied from 5 to 0.1 mol% to observe the lower limit of alkaline-earth salt composition required to deposit Sr or Ba. The results of this work suggest that alkaline-earth fission products accumulated in molten salts (Sr^{2+} and Ba^{2+}) can be recovered into liquid Bi by electrochemical separation, which could be employed as a critical step for recycling the process salt (LiCl-KCl) in order to minimize the generation of additional nuclear wastes.

525498 Electrochemical Separation of Cs from Molten Salts Using Liquid Metal Electrodes

Abstract:

The feasibility of removing Cs from LiCl-KCl-CsCl (eutectic-3 mol%) using liquid Bi, Sn, and Pb electrodes was examined via thermodynamic properties measurements as well as electrochemical depositions. The accumulation of Cs^{137} in the LiCl-KCl-UCl_3 electrorefiner electrolyte leads to frequent replacement and disposal as high level nuclear waste due to the high heat density and short half life of Cs^{137}. Unfortunately, in the chloride system Li and K are less noble than Cs, making Cs difficult to remove from a LiCl-KCl-based electrolyte. However, by leveraging the strong atomic interactions between Cs and liquid metals such as Bi, Sn, and Pb, we examine the possibility of selectively removing Cs from chloride-based molten salt, providing a pathway to reducing this particular source of nuclear waste. Electrochemical depositions on these liquid metal working electrodes were performed up to a depth of 100 C g^{-1}, resulting in co-deposition of Cs and Li up to (4 mol% Cs - 12 mol% Li) at 500 °C in Bi. Electrochemical deposition experiments were complimented by electromotive force measurements for the Cs-Bi system to elucidate the thermodynamic properties (namely, the activity of Cs in Bi) leading to the observed shift in expected deposition behavior due to the liquid metal electrode. This work suggests the possibility of using liquid metal electrodes to remove Cs from molten salts and present a potential path towards reducing the volume of nuclear waste generated from the electrorefining process.

527292 Electrochemical Analysis of Bi-Analyte Electrorefiner Salt with High Concentration of UCl_3

Abstract:

During the spent nuclear fuel electrorefining process, the spent fuel is oxidized into the molten salt mixture followed by selective reduction on a cathode. Consequently, a large inventory of fissile actinides is held up in the salt. Thus, an important requirement for operating the process is to keep track of the actinides in the salt mixture to prevent proliferation. Several recent publications on either low UCl_3 concentration or single analytes show that electrochemical senor could potentially monitor salt concentration in real time. However, the UCl_3 concentration in the electrorefiner is often between 5-10 wt% during operation. Therefore, the objective of this research is to focus on bi-analyte systems with up to 10 wt% UCl_3. Two bi-analyte systems U/Mg and U/Gd were studied. Mg is a surrogate for Pu due to their similar reduction potentials, and Gd is selected to represent the rare earth metals in fission products. In this research, normal pulse voltammetry (NPV) was selected to prevent working electrode surface area growth caused by U deposition. Optimizations were made to the procedure to make NPV compatible and highly accurate for this application. Through the study of U/Mg system, NPV parameters were established, and it was learned that migration currents are comparable in magnitude to diffusional currents. Solution resistance compensation and small working electrode were applied for further optimization during the study of U/Gd system. The final concentration prediction average error values for UCl_3 and $GdCl_3$ are 1.6% and 2.70% respectively. This result shows that NPV is a suitable method for monitoring spent nuclear fuel electrorefiner molten salt metal concentrations in real time. It is also a promising approach for monitoring actinide concentrations in molten salt reactors.

528296 Monitoring of Actinide Concentrations in Molten LiCl-KCl Salt Using Alpha Spectroscopy

Abstract:

In order to develop a sensor for measuring actinide concentrations in molten electrorefiner salt, a process for electrodeposition of U and Th from LiCl-KCl-UCl_3-$ThCl_4$ onto Pt-coated surfaces has been developed. To facilitate the counting of alpha particles by the sensor, the objective is to electrodeposit a uniform, one micron thick layer of actinides. Such a deposit was accomplished using repeating chronoamperometry (RCA) tests in LiCl-KCl-UCl_3, LiCl-KCl-$ThCl_4$, and LiCl-KCl-UCl_3-$ThCl_4$ molten salt mixtures at 500°C. Uniformity of the deposits was verified using scanning electron microscopy with energy dispersive x-ray spectroscopy (SEM-EDXS). Time to pass a fixed amount of electric charge through the cell was correlated with concentrations of UCl_3 and $ThCl_4$ in the salt mixtures. As expected, the deposition time was inversely proportional to actinide concentration in the salt. The actinides could be stripped and re-deposited multiple times on a given Pt-coated coupon with repeatable time results. This method shows great promise for simultaneously measuring concentrations of multiple actinides with very close electrochemical reduction potentials. It thus offers potential functionality beyond what is possible with only electrochemical methods such as cyclic or normal pulse voltammetry.

532816 Thermodynamic Properties of Strontium-Lead Alloys Determined By Electromotive Force Measurements

Abstract:

During electrorefining of metallic used nuclear fuel, alkaline-earth fission products (Sr and Ba) are oxidized at the anode and accumulate in the molten salt electrolyte. To evaluate the viability of electrochemically separating these stable ions from the electrolyte using a liquid Pb cathode, the thermodynamic properties of Sr-Pb alloys were investigated by electromotive force (emf) measurements. A Sr(s)|CaF_2-SrF electrochemical cell using a solid state electrolyte (binary CaF_2-SrF emf values at 773 K-1073 K for ten Sr-Pb alloys at mole fractions $0.07 \leq x_{Sr} \leq 0.45$. These emf measurements were used to determine thermodynamic properties, including activity, partial molar enthalpy, and partial molar entropy. At 873 K, activity values of Sr in Pb were as low as 1.69×10^{-9} at mole fraction $x_{Sr} = 0.07$, implying strong atomic interactions between Sr and Pb. Phase characterization utilizing X-ray diffraction (XRD) was conducted on alloy compositions $0.07 \leq x_{Sr} \leq 0.75$ and thermal analysis was conducted on compositions up to $x_{Sr} = 0.34$ using differential scanning calorimetry (DSC). Combining phase analysis and transition temperatures, the Sr-Pb phase diagram was refined. Based on the measured thermodynamic properties, the electrochemical deposition into a liquid Pb cathode was conducted in molten LiCl-KCl-$SrCl_2$ at 500°C and the Pb cathode was characterized to examine the deposition products (e.g., Sr, Li, and K).

523953 Evaluation of Concentrations of Residual Water, Hydroxides, and Oxides in Molten Anhydrous $CaCl_2$

Abstract:

Molten calcium chloride can be used as an electrolyte for reduction and electrorefining of actinides. However, its hygroscopic nature may present a barrier to this application. Simply heating the salt to its molten state may not entirely remove the water. At elevated temperature, the water may hydrolyze $CaCl_2$ to form calcium hydroxide or oxideâ€"both of which are expected to have substantial solubility in the molten salt. Such impurities are expected to interfere with the designed oxidation/reduction reactions in the electrorefiner. In this study, we examined a number of methods for quantifying these contaminantsâ€"including thermogravimetric analysis, cyclic voltammetry, and acid-base titration. The starting material was commercially prepared ultradry $CaCl_2$, which was melted to 850Â°C. Thermogravimetric analysis of this salt indicated a 1 wt% mass loss on heating to 770Â°C, indicating a water content of at least that level. Cyclic voltammetry of the as-received $CaCl_2$ indicated the presence of a reduction peak that could be attributed to calcium hydroxide. To verify this assumption, current peak response to known additions of calcium hydroxide was recorded. The resulting correlation can be used to make in-situ measurements of calcium hydroxide concentration in molten $CaCl_2$. Acid-base titration yields the total concentration of $Ca(OH)_2$ and CaO in solidified samples of the salt. The combination of these three methods can be used to determine removable residual water as well as hydroxides and oxides that form from non-removable water. This presents a route for future work in $CaCl_2$ molten salts towards developing an effective drying process.

538889 A Lightweight Betaemitter for Power Applications

Abstract:

Betavoltaic batteries have been around for quite some time but have yet to be manufactured with enough efficiency to safely produce an equivalent battery to current low power batteries. A betavoltaic battery produced with tritium has the potential to produce a battery with a life of around 12 years. Lithium can be easily intercalated into highly-oriented pyrolytic graphite (HOPG). The trapped lithium can then be neutron-irradiated to form tritium. This would produce a HOPG structure with tritium infused throughout. Understanding the lithium-intercalated HOPG structure is important to determining what might happen upon irradiation. Different methods of analysis were used for characterizing the lithium intercalated HOPG. Auger electron spectroscopy was used to determine if lithium was present. X-ray diffraction determined the spacing between the graphene planes. Although scanning electron microscopy proved inconclusive, transmission electron microscopy and diffraction provide valuable information about the lattice.

539820 Adsorption and Filtration of Lignocellulosic Hydrolyzates Using Fibrous Depth Filters

Abstract:

Adsorption and Filtration of Lignocellulosic Hydrolyzates Using Fibrous Depth Filters

Mary Jennifer Puthota, Thomas Dwyer Stuart and Bandaru V. Ramarao

Empire State Paper Research Institute

Department of Paper and Bioprocess Engineering

State University of New York,

College of Environmental science and Forestry,

1 Forestry Drive, Syracuse, New York 13210, USA

Abstract

Lignocellulosic hydrolyzates contain lignin in dissolved and colloidal form which must be separated before fermentation to prevent inhibition of bioconversion of the sugars. The most common detoxification techniques to remove lignin involve acidification and precipitation followed by adsorption using activated carbons.

Depth filtration has emerged as a useful technology in Bioseparations to purify solutions containing colloidal particles and macromolecular species. Lignocellulosic hydrolyzates can be purified using fibrous depth filters effectively, especially by integrating a flocculation step ahead of depth filtration.

In this paper, the application of fibrous depth filters using unbleached and bleached kraft pulp fibers (hardwood and softwood) is presented. Since fibers are anionic, as are the lignin particles in hydrolyzates, it is necessary to cationize the fibers using polyelectrolytes such as poly DADMAC to cause filtration. Another route is the use of PEO polymers to attach the lignin particles to the fibers.

Both of these techniques are considered and preliminary experimental results on separations are presented. Mathematical models for depth filters of composed of such fibers are presented and analyzed to show the impact of different operating variables.

539053 Process Modeling, Simulation and Optimization of Process Intensification in Integrated Biorefineries

Abstract:

Lignocellulosic feedstock as a renewable resource for biofuels, bioenergy and value added, renewable bioproducts is gaining increasing attention throughout the world. First generation biofuels from corn and sugar and second generation biofuels from cellulosic biomass are excellent examples of emerging biorefineries. Integrated biorefineries are an emerging concept using thermochemical, biological and biochemical approaches to conversion of the lignocellulosic feedstocks to multiple products including biofuels, bioenergy, biochemicals and value-added renewable bioproducts. Butanol is not only a good drop-in fuels with higher energy density than ethanol, but also a very important chemical.

Butanol can be produced by fermentation of sugars derived from lignocellulosic biomass. However, there are key challenges in butanol biorefineries: (i) severe product (i.e butanol) inhibition due to toxicity to the microorganisms during fermentation, which leads to low butanol yield and productivity, and (ii) expensive downstream processing (product separation and purification). This paper is focused on the process intensification and integration of bioprocessing and separation technologies, i.e., integration of efficient separation process with fermentation for in situ removal of product inhibitor to increase the butanol yield and productivity and lower the production cost. The comparison of energy consumption between different separation approaches is carried out by modeling and analysis of the integrated biorefinery, based on recent research, with process intensification and integration of bioprocessing and separation of products (e.g., Fig 1). Process optimization will also be carried out.

533901 Process Intensification of the Hydrolysis of Cellulosic Fibers By Integration of Membrane Separations and Hydrolysis for Enzyme Recycle

Abstract:

Waste fibers from papermaking are often rejected due to quality and production issues. Since the rejected fibers also known as 'fines' are discharged as solid waste placing significant strain on landfills, bioconversion into sugars by hydrolysis of their cellulosic content can be very attractive to the paper industry. The present work presents a method for process intensification of the hydrolysis of waste fines by integrating membrane separation for enzymes and recycle along with hydrolysis in a separation hydrolysis reactor. The performance of the reactor is analyzed and compared with experimental data on hydrolysis.

515606 A Low Energy Separation Process to Isolate Xylose from Aqueous Hydrolyzate Streams

Abstract:

Water removal is a key process step in a majority of process industries including biomass conversion and biofuels industry. Unit operations such as evaporation, concentration, membrane filtration, etc., are widely used for water removal; however, they contribute significantly to the overall processing cost to produce biofuels primarily due to their high energy requirement. For example, it takes 9,274 BTU to remove 1 gal of water from 70 F via evaporation. Sugar extraction from biomass and its conversion to bioproducts via fermentation is typically a water based process â€" requiring large volumes of water removal to produce the products. Moreover, the products resulting from fermentation are typically low energy density fuel alcohols, organic acids, etc. If the sugars extracted can be made available in dry solid form, then the sugars can be converted to high energy density biofuels and bio-hydrocarbons via chemical synthesis. Conn Center at University of Louisville developed a highly energy efficient separation process for hemicellulose (C5) based sugars, specifically xylose, which is conducted under ambient conditions and delivers xylose in a dry form from an aqueous hydrolyzate stream containing xylose.

A boron chelation chemistry is used in this xylose isolation process. This chemistry and the associated methodology can be extended to the recovery of anhydrosugars. The current commercially practiced xylose separation process involves complex chromatographic separation and high-energy evaporation steps and generates large volumes of waste streams. The boron chelation chemistry can clearly lead to process intensification: it is done under ambient conditions, consists of fewer number of process steps, and leads to minimal waste generation. The isolation scheme also includes recovery of boron for reuse in the front end of the process.

524672 Organotemplate-Free Beta Zeolites: From Zeolite Synthesis to Hierarchical Structure Creation

Abstract:

Zeolite beta is a large-pore, crystalline material that features an intergrowth of two or more polymorphs comprising a three-dimensional, 12-membered ring channel system with great potential in refinery, petrochemical, and chemical applications. During the past decade, the seed-directed organotemplate-free strategy has been successfully developed for the synthesis of beta zeolites. This presentation will focus on extending the organotemplate-free strategy to the synthesis of hierarchical beta zeolites without using organic structure-directing agents (OSDA), pore-directing agents (PDA), or mesoporous templates, which is achieved by hydrothermal desilication that is amenable for beta crystallization. Organotemplate-free creation of hierarchical structures will be demonstrated for both OSDA-free and conventional beta zeolites as parent zeolites. The resulting hierarchical beta zeolites have higher pore volumes and exhibit improved catalytic performance for the conversion of bulky molecules. They show promise for a variety of applications in which the presence of large mesopores are required to reduce diffusion limitation and coking formation.

535471 Tuning the Molecular Design of Tertiary Amine Catalysts on Amorphous Mesoporous Silica Supports for Selective Glucose Isomerization and Acid-Base Cooperative Reactions

Abstract:

Amorphous silica materials offer a tunable, high surface area platform for creating heterogeneous catalysts. While mesoporous silica supports such as SBA-15 are presented as an idealized structure, real catalytic materials are quite complex and can have molecular interactions that facilitate some reactions while inhibiting others. The key challenge is understanding how these molecular interactions affect catalytic performance. Through combining molecular synthesis, advanced spectroscopy, and computational work, molecular interactions that occur in amorphous mesoporous silica are elucidated. Specifically, we investigate amorphous silica materials containing surface silanols that are functionalized through grafting organosilanes containing amines. We identify that the amine structure and loading as well as the catalyst structure strongly impact observed amine-silanol interactions. The observed amine-silanol interactions have implications for catalytic behavior observed for glucose isomerization and the Knoevenagel reaction. Through tuning the overall catalyst design, the overall catalyst activity and selectivity can be improved. Overall, these results provide unique insights on the nature of the surface interactions of the catalytic sites with the silica surface that improve our understanding of these materials.

538241 Template Free Synthesis of Palladium Immobilized Ordered Mesoporous Resin for Drug Synthesis on a Chip

Abstract:

Packed bed microreactors are strong candidates for pharmaceutical applications to efficiently screen new catalyst and synthesize new drugs. In this project we synthesized a novel template-free ordered mesoporous resin (OMR) immobilized with palladium. The primary catalyst characterization showed high BET surface area (675.34 m2/g) and high thermal stability (TG more than 255 °C) which made it a superior heterogeneous catalyst for cross-coupling reaction. The batch experiment results showed high conversion rate (98%) for Suzuki-Miyaura cross-coupling reaction up to 5 times of recyclability. A packed bed microreactor was developed to investigate the catalyst efficiency for cross-coupling reaction. The microreactor employs a U-shape channel which is filled with Pd-OMR catalyst. The reaction carried out across a range of liquid flow (0.1 t0 0.5 μl/min) rates, bed porosity percentage, and temperature (65 t0 85 °C). Experiments seek to assign a reaction conversion and catalyst stability to each flow rate. The results show that the conversion in the continuous system is greater than the batch system, which is due to the mass transfer enhancement induced by the porosity of the catalyst. Reaction time within the microchannel is sufficiently small so that selectivity towards the product is nearly 100% across all flow regimes. In this work, we relate our map of conversion to the flow behavior and microreactor geometry, and we discuss steps for further exploring the mass transfer characteristics of the flow. This heterogeneous reactor architecture shows a promising pathway synthesize new drugs.

538316 Mesoporous Zeolites Produced By Solid Crystallization and Their Hydrogenation Properties

Abstract:

Hydrogenation of aromatics is one of the most important unit operations in modern oil-refinery and coal liquefaction industries. The produced cycloalkane can significantly increase the cetane number of fuel, improve its combustion performance, and enhance its thermal stability even in some critical environments. Some recent studies found that when supported on acidic materials such as USY or Beta zeolites, noble metal catalyst such as platinum (Pt) showed much improved catalytic activity and sulfur tolerance in deep hydrogenation of aromatics. However, the sole microporosity of these zeolites often makes these reactions less effective when bulky aromatic hydrocarbons such as naphthalene (a major compound of light cycle oil or coal tar) are involved due to the significant mass transfer limitation.

Here we investigate the naphthalene hydrogenation performance over platinum loaded on new mesoporous zeolites (Meso-ZSM-5). A new solid-state crystallization approach is developed to synthesize mesoporous zeolites at ambient pressure without addition of water or other solvents. The adjacent nanocrystals are packed and eventually joined into a large monolithic zeolite, creating many regular inter-lattice space to serve as mesoporous pathways for better mass transport and acidic site utilization. The hydrogenation was performed between 150-280°C with a hydrogen pressure of 2-8 MPa. Compared to the hydrogenation over Pt/microporous ZSM-5 and Pt/Al2O3, Pt/Meso-ZSM-5 shows better conversion and high selectivity in the absence and presence of dibenzothiophene (DBT). It excellent catalytic performance and sulfur tolerance is further tailored with the acidity and mesoporous structure of Meso-ZSM-5.

538444 Pillared Two-Dimensional Titanium Silicalite-1 Zeolite: Synthesis, Characterization and Catalytic Applications

Abstract:

Derived from the two-dimensional (2D) zeolite precursors, pillared zeolites contain mesopores between zeolitic layers and micropore within each zeolite layer. Thus, pillared zeolites couple intrinsic catalytic activity of micropores and fast mass transport of mesopores in one single materials, offering enhancement in catalyst utilization during catalytic processing of heavy and bulky reactant molecules. In this talk, we highlight our recent work on development of pillared 2D titanium silicalite-1 (TS-1) zeolites. The synthesis of TS-1 precursors, pillaring of the precursors into pillared 2D TS-1 zeolites, identification of acid site locations, textural property characterization, as well as catalytic applications in phenol hydroxylation and bulky molecule epoxidation reactions were studied sequentially and systematically. The present work brings a new addition to the structure and composition modifications of 2D lamellar zeolite with more open structures and diverse active sites for processing bulky molecules.

530660 Synthesis Methods to Influence Framework Al Arrangements in CHA Zeolites and Consequences for NO_x Selective Catalytic Reduction

Abstract:

The use of organic and inorganic cations as structure-directing agents (SDAs) in zeolite crystallization provides a mechanism to influence the arrangement of anionic framework Al centers (Al−O(−Si−O)x−Al) in isolated ($x \geq 3$) or paired ($x = 1, 2$) configurations. We discuss this within the context of chabazite (CHA), a high-symmetry framework comprised of one crystallographically-distinct T-site, which facilitates complementary experimental and theoretical approaches to probe how the molecular and electronic structure of SDAs influence Al arrangement. Titration protocols to quantify the number of paired Al sites in CHA zeolites were developed using divalent Co^{2+} and Cu^{2+} cations, which were validated by measuring UV-Visible spectra and quantifying residual protons that remain after cation exchange. These experimental characterizations are consistent with calculations of divalent cation siting energies at different framework Al arrangements in the CHA framework. CHA zeolites (Si/Al = 15) crystallized from mixtures containing only low charge-density organic N,N,N-trimethyl-1-adamantylammonium cations ($TMAda^+$) as the SDA contain predominantly isolated Al sites, because adamantyl groups (~0.7 nm diam.) impose steric constraints that limit the occupation of each CHA cage (~0.7 nm diam.) to only one $TMAda^+$ molecule. The partial replacement of $TMAda^+$ with high charge-density Na^+ cations in crystallization media, holding other variables constant, results in the formation of paired Al sites in amounts that correlate with the amount of Na^+ incorporated within crystalline CHA products. This cooperation between Na^+ and $TMAda^+$ during CHA crystallization reflects incorporation of Na^+ near the quaternary ammonium group of a $TMAda^+$ molecule occluded within the CHA cage, which preserves dispersive contacts between non-polar siloxane regions of the zeolite lattice and the hydrophobic adamantyl group of $TMAda^+$. Density functional theory calculations, performed on a 36 T-atom CHA unit cell (Si/Al = 17), were used to compute the energies of different Na^+ and H^+ configurations as a function of Al-Al proximity. DFT-derived energies indicate that Na^+ siting is preferred when two Al are located in the same six-membered ring, suggesting that the occlusion of Na^+ during CHA crystallization preferentially directs the formation of paired Al sites. We also discuss our recent work combining experiment and theory to develop alternate crystallization routes using mixtures of other SDAs to influence framework Al arrangements in CHA, and discuss the consequences of these site distributions for NOx selective catalytic reduction (SCR) with ammonia. These findings provide further insight into the cooperative and competitive interactions between organic and inorganic SDAs on the arrangement of Al atoms in CHA zeolites.

522958 Unraveling and Tuning Surface and Catalytic Chemistry of Zr_6O_8Â NodesÂ in Metal Organic Frameworks

Abstract:

The nodes of some metal-organic frameworks (MOFs) closely resemble small pieces of metal oxides, exemplified by the $[Zr_6O_4(OH)_4]^{12+}$ nodes of UiO-66, UiO-67, and NU-1000. Defects including missing linkers or structural vacancies on these nodes are sites on which many acid-base-catalyzed reactions occur, exemplified by alcohol dehydration. These sites are also anchoring sites for metal complexes. As these MOFs are synthesized, various ligands form on the nodes, and their identification has taken years to resolve. The resolution has emerged from characterization of the MOFs by infrared and nuclear magnetic resonance spectroscopies. The initially present ligands were determined to be formate, when HCl or formic acid is used as a modulator in the synthesis, and acetate, when acetic acid is used as a modulator. These ligands can be replaced by alkoxy ligands by treatment of the MOF with an alcohol, and the alkoxy ligands can be converted to hydroxy when the MOF is further treated with water. The nodes were tested as catalysts for ethanol dehydration, which takes place selectively to make diethyl ether but not ethylene at 473-523 K. Calculations by density functional theory have shown that the key to the selective catalysis is the breaking of node-linker bonds or the unintended adjacency of open sites resulting from the synthesis to allow catalytically fruitful bonding of reactant ethanol to neighboring sites on the nodes and thereby facilitate the bimolecular ether formation reaction through an S_N2 mechanism. At the catalyst functions, ethanol facilitates the breaking of node-linker bonds by formation of esters that lead to the destruction of the MOFs. Thus, the choice of the linkers and the synthesis conditions is motivated by the goal of optimizing the activity and stability of the MOF as a catalyst. We address this issue and further the issue of how the linkers affect not only the stability but also the reactivity and catalytic activity of the MOF nodes. Because the linkers are ligands bonded to the nodes, they affect the electronic properties of the nodes. We used μ_3-OH groups on the nodes as sensitive indicators of the electronic properties of the node, characterizing them by infrared spectroscopy and catalyst performance. Thus, we present the first experimental evidence showing how linker ligands affect the catalytic properties of these nodes.

528928 Rapid and Low Temperature (< 80 °C) Hydrotrope Chemistry for Economic and Sustainable Production of Sugar/Biofuels, and Lignocellulosic Nanomaterials

Abstract:

Here we report a rapid and low temperature (≤ 80°C) wood fractionation process for producing wood pulp, cellulose and lignin nanomaterials, and sugar/biofuels. Approximately 90% of wood lignin can be dissolved at 80°C for 20 min. Equivalent delignification using alkaline pulping can be achieved only at 170°C for 2 h. The novel process fractionated wood into two fractions: (1) a primarily carbohydrate-rich water-insoluble solid fraction that can be used for sustainable production of fibers such as dissolving pulp, lignocellulosic nanomaterials, and/or sugars through subsequent enzymatic hydrolysis, and (2) a spent acid liquor stream containing mainly dissolved lignin that can be easily precipitated as lignin nanoparticles simply by diluting the spent acid liquor. The dissolved hemicellulose sugars can be used to produce furfural after lignin precipitation. The chemical used has a low water solubility, which facilitates efficient recovery simply using commercially proven crystallization technology by cooling concentrated spent acid solution to ambient temperatures to achieve environmental sustainability.

534186 HPLC/Qtof-MS for Analysis of Oligomeric Compounds (pyrolytic lignin) from Crude Bio-Oils and Products from Pyrolysis of Model Compounds

Abstract:

Fast pyrolysis of biomass has become a popular way to obtain liquid products with complex mixtures, including medium-polar and polar monomers, water, and oligomeric compounds (or pyrolytic lignin, PL). Among these, the oligomeric fraction, accounted for about 20 wt.%, have negative effects on bio-oil and are difficult for further utilization. Therefore, it is of vital importance to understand the structure and formation of pyrolysis lignin when attempting to make full use of bio-oil.

In this work, pyrolysis bio-oil derived from three kinds of lignocellulosic biomass (Rice Straw, RS; Cotton Stem, CS; Walnut Shell, WS) were used to obtain PL samples (simplified as RS-PL, CS-PL, WS-PL) as fine homogeneous powder by precipitation method. Besides, aromatic monomers (guaiacol, eugenol, vanillin) and acetic acid were mixed together as model compounds of bio-oil vapor for fast pyrolysis at 550 to simulate the actual pyrolysis condition to study the formation of pyrolytic lignin. The products were collected as model compounds pyrolysis sample (M-PL). All the samples mentioned above were dissolved in methane for further analysis.

Since high-performance liquid chromatography coupled with UV detector and quadrupole-time-of-flight mass spectrometry (HPLC/QTOF-MS) is a powerful tool for separation and structural analysis of big unknown molecules in complex mixtures, it is used here to obtain molecular weight and structure information on the major components of the three PL samples and model compounds pyrolysis sample.

The results show that the UV chromatograms and total ion chromatograms (TIC) of the three PL samples have high similarity on distribution and intensity of chromatographic peaks, indicating a lot of compounds in PL samples are similar. Combined with TOF MS spectra and TOF MS^2 spectra, molecular formula and possible structure of the common components are confirmed, such as $C_{19}H_{18}O_4$, $C_{18}H_{18}O_4$, $C_{17}H_{16}O_4$, $C_{16}H_{14}O_4$, $C_{13}H_{12}O_2$, $C_{12}H_{10}O_2$, $C_9H_{12}O_2$, $C_8H_{10}O_2$. They are mostly aromatic dimers and aromatic monomers. RC-PL sample was then taken as a representative to compare with M-PL sample and some similarities in UV and TIC chromatograms were found as well. According to extra ion chromatogram (XIC), aromatic monomers ($100 \le Mm \le 180$, $4 \le DBE < 8$) of the two samples match really well, and a certain number of aromatic dimers or trimers ($180 < Mm \le 400$, $DBE \ge 8$) with high intensity are also found to be the same. The results indicate that the oligomeric fraction of crude bio-oil may form from aromatic monomers through free-radical polymerization during the pyrolysis process. Furthermore, seven compounds are proposed to be considered as typical components in oligomers fraction. The molecular formulas (m/z, DBP) are $C_{16}H_{14}O_2$ (237.0921, 10.0), $C_{14}H_{12}O_2$ (211.0765, 9.0), $C_{14}H_{10}O_3$ (225.0557, 10.0), $C_{13}H_{12}O_2$ (199.0765, 8.0), $C_{13}H_{10}O_2$ (197.0608, 9.0), $C_{12}H_{10}O_2$ (185.0608, 8.0), $C_{12}H_8O_2$ (183.0452, 9.0).

523257 Steam Reforming of Toluene and Biomass Tar over Biochar Supported Ni Nanoparticles: Effects of Ni Particle Size on Catalytic Activity and Stability

Abstract:

A series of biochar supported Ni catalysts (Ni/BC) were explored for steam reforming of toluene, with an emphasis on the effects of Ni particle size on the product distribution and catalytic stability. Metallic Ni nanoparticles were formed upon pyrolysis treatment of impregnated biomass via carbothermal reduction. The influence of preparation parameters (Ni loading amount and pyrolysis temperature) on the Ni particle size was examined. Ni/BC pyrolyzed at 600 °C with a Ni loading of 5% (5Ni-600/BC) displayed higher specific surface areas and higher metal dispersion (i.e. lower particle size) than other catalysts. In addition, 5Ni-600/BC showed superior catalytic performance in terms of initial activity (TOF) and stability over other Ni/BC catalysts, indicating the positive role of small particles with more corner and edge sites. Two conventional metal oxides supported Ni catalysts (Ni/ZrO_2 and Ni/Î^3-Al_2O_3) were selected for comparison with 5Ni-600/BC in long-term stability tests. Although characterization of the spent catalysts revealed that 5Ni-600/BC experienced severe coke deposition and metal sintering, it performed more stable than the reference catalysts, which can be attributed to its significantly larger specific surface area available for coke tolerance. 5Ni-600/BC was further examined for its performance in steam reforming of real biomass tar. Compared with BC alone, Ni/BC not only reduced the overall amount of tar, but also cracked some large molecules into smaller ones. Our study demonstrated that Ni/BC is a cheap and easy-to-prepare catalyst with promising prospects for toluene/tar reforming. However, the catalyst stability still needs to be improved by inhibiting metal sintering and coke deposition in future work.

536816 Optimize Fermentation Conditions By Using Glucose As a Substrate to Produce (R)-3-Hydroxybutyric Acid with *Burkholderia Cepacia*

Abstract:

Polyhydroxyalkanoates (PHAs) are degraded as promising polymers to substitute common plastics, the studies about PHAs have become more and more popular. Poly-(R)-3-hydroxybutyrate (PHB) is one of these attractive and commercially manufactured polyhydroxyalkanoates. (R)-3-hydroxybutyric acid (3-HBA) is a valuable monomer of PHB, which has wide potentials to be applied in industries and medical treatments. We use glucose as a substrate to optimize some fermentation conditions, such as temperature, to increase the concentration of 3-HBA during the batch fermentations with *Burkholderia cepacia*.

529878 Organosolv Pretreatment of Hybrid Pennisetum for the Production of Lignin and Enzymatically Digestible Cellulose

Abstract:

Four different types of solvents were evaluated for organosolv pretreatment with hybrid pennisetum. Under lower temperature (100 degrees C, 50mM H2SO4 as catalyst), the yields of delignification with acetone, ethanol, GVL and tetrahydrofurfuryl alcohol were 32.8%, 37.6%, 43.2% and 50.2%, respectively, and which results were consistent with the predicts of Hansen solubility parameters. Moreover, a parametric study using tetrahydrofurfuryl alcohol and focusing on acid catalyst dose, temperature, reaction time and solid-liquid ratio were performed to determine their influence on delignification, xylan hydrolysis, and enzymatic cellulose digestibility. Major process parameters were found to be temperature and acid dose. Optimisation of the process towards delignification resulted in a maximum lignin removal rate was 98.1% under 120 degrees C, 50mM H2SO4 as catalyst, solid-liquid ratio 1:12 and reaction time 2 h, and the cellulose retention rate was 86.1%. The digestibility of cellulose rich residues after pretreatment was approximately 99.0%, which was 2.6 times as much as that of the untreated raw material. The composition and structure of the raw materials, the pretreated residues and recovered lignin were analyzed by SEM, XRD, FT-IR, TGA and HSQC. The results indicated that the pretreatment of organosolvs could enhance the enzymatic hydrolysis of lignocellulosic by effectively removing lignin and hemicellulose, then destroying smooth surface structure of raw materials. And the characterization results of recovered lignin indicate that the lignin molecules have undergone significant bond breakage during pretreatment.

KEYWORDS:

Organosolv pretreatment; Enzymatic hydrolysis; Recovered lignin

532101 Dual Functional Zr-KIT-5 Shows Remarkable Activity for Depolymerization of Corn Stover Lignin into Stable Phenolic Monomers

Abstract:

Dual Functional Zr-KIT-5 Shows Remarkable Activity for Depolymerization of Corn Stover Lignin into Stable Phenolic Monomers

Kakasaheb Y. Nandiwale[1,2], Andrew M. Danby[1], Anand Ramanathan[1], Raghunath V. Chaudhari[1,2], and Bala Subramaniam[1,2]

[1]Center for Environmentally Beneficial Catalysis, University of Kansas, Lawrence, Lawrence, Kansas 66047, United States

[2]Department of Chemical and Petroleum Engineering, University of Kansas, Lawrence, Kansas 66045, United States.

Keywords: lignin depolymerization, Zr-KIT-5, Lewis acid, acetic acid

We previously showed that Zr-KIT-5 exhibits superior performance than H-ZSM-5 for depolymerization of dealkaline lignin yielding up to 77wt% of THF-soluble products (monomers and dimers). Structural degradation and loss of acidity occur in spent Zr-KIT-5 catalyst primarily due to a lack of hydrothermal stability in aqueous methanol solvent.[1]

These insights motivated us to screen various non-aqueous solvents such as 1,4-dioxane, acetic acid, and Î³-valerolactone (GVL) for their ability to solubilize lignin and to avoid the hydrothermal degradation of Zr-KIT-5 catalyst. As expected, the organosolv (corn stover-derived) lignin was completely soluble in acetic acid, the solvent used for separating the lignin from cellulose fraction. Significantly, the solvent exhibited thermal stability at typical depolymerization temperatures. Furthermore, Zr-KIT-5 catalyst showed much better physical stability in acetic acid solvent compared to aqueous (water + methanol) solvent. About 12 different phenolic monomers (including a few that are acetylated) were identified in the depolymerized product mixture by GC-MS/FID. The Lewis-acid sites of Zr-KIT-5 catalyst appear to facilitate not only depolymerization of lignin, but also acetylation of the monomers, thus preventing monomeric condensation/repolymerization. High yield of identified monomers (28 wt%), comparable to those reported to date, and ethyl acetate soluble aromatic species (43 wt%) were obtained with thermo-solvolytic treatment of lignin followed by catalytic depolymerization over Zr-KIT-5 catalyst. The Zr-KIT-5 was found to be a promising solid Lewis acid catalyst displaying stable activity for up to five catalytic lignin depolymerization cycles.

Reference:

(1) K.Y. Nandiwale, A.M. Danby, A. Ramanathan, R.V. Chaudhari, B. Subramaniam, *ACS Sustainable Chem. Eng.*

2017, *5*, 7155.

525859 Reaction Condition Optimization for the Scalability of 1,4-Anhydroerythritol and Xylitol Conversion Via Heterogeneous ReO_X-Pd/CeO_2 Catalysis

Abstract:

Some biomass is rich in sugars that contain hydroxyl groups, which can either undergo deoxydehydration to form double bonds or hydrodeoxygenation (HDO) to remove vicinal hydroxyl groups. HDO via deoxydehydration followed by a hydrogenation step allows for the simultaneous removal of multiple vicinal hydroxyl groups[1]. Literature has shown that ReOx-Pd/CeO_2 is a better catalyst for the simultaneous HDO of 1,4-anhydroerythritol as compared to ReOx-Pd on other oxide supports[2]. The simultaneous HDO of 1,4-anhydroerythritol, which produces tetrahydrofuran, was used as a model reaction for HDO. Pressure and temperature effects on the system will be explored through a design of experiment in order to elucidate scalable conditions for HDO reactions. Xylitol is another biomass derived chemical that can be upgraded to value-added products through simultaneous HDO, but there is a lack of understanding on how ReOx-Pd/CeO_2 catalysts perform for this reaction[2]. Thus reaction condition optimization could benefit this process significantly. In order to obtain a detailed understanding of the optimal conditions for these reactions, a design of experiment was enacted to discern the effects and interactions of temperature, pressure, and catalyst loading. Xylitol was chosen as a reactant to compare to the model 1,4-anhydroerithritol reaction due to it having three vicinal cis-hydroxyl groups and in order to form 1,2,5-pentanetriol and other products by selectively removing hydroxyl groups. The xylitol HDO has been conducted in literature at 80bar and 443K[2]. Reducing the pressure would allow for this process to become more economically favorable and scalable. The findings from the 1,4-anhydroerythritol reaction design of experiments were extended to the xylitol reaction in order to find trends between the operating conditions, resulting in more reasonable operating conditions.

References

(1) Ota, N.; Tamura, M.; Nakagawa, Y.; Okumura, K.; Tomishige, K. Performance, Structure, and Mechanism of ReO $_x$ â€"Pd/CeO $_2$ Catalyst for Simultaneous Removal of Vicinal OH Groups with H $_2$. *ACS Catal.* **2016**, *6* (5), 3213â €"3226.

(2) Ota, N.; Tamura, M.; Nakagawa, Y.; Okumura, K.; Tomishige, K. Hydrodeoxygenation of Vicinal OH Groups over Heterogeneous Rhenium Catalyst Promoted by Palladium and Ceria Support. *Angew. Chemie - Int. Ed.* **2015**, *54* (6), 1897â€"1900.

527496 Selective Tuning of the Glycerol C-O Bond Cleavage Sequence on Copper-Modified Molybdenum Carbide Surfaces

Abstract:

Biodiesel is an important renewable resource for addressing energy and environmental challenges of current society[1,2]. Glycerol is a major byproduct of biodiesel production[3] and can be upgraded into value-added products via the selective hydrodeoxygenation reaction. In this work, we demonstrated the selective tuning of the C-O bond cleavage sequence of glycerol by varying the Cu coverage on Mo_2C. Temperature Programmed Desorption (TPD) results show that the Mo_2C surface cleaves all the C-O bonds of glycerol to produce propylene; the Cu/Mo_2C interface tends to cleave a primary and a secondary C-O bond of glycerol to form allyl alcohol; the Cu surface prefers to break a primary C-O bond to form acetol. Surface vibrational studies indicate that glycerol undergoes initial O-H scission on Mo_2C and the C-O bonds are weakened due to the strong interaction between the oxygen in glycerol and Mo_2C. As Cu coverage increases, the O-H bonds of glycerol are preserved, and the O-Mo interaction is reduced. Activation barrier calculations from DFT further reveal the different bond-breaking capability of the Mo site, Mo-Cu interface and Cu site on Cu/Mo_2C due to their different oxygen binding energies. The combined model surface experiments and DFT calculations suggest the feasibility of selectively controlling the C-O bond breaking sequence of glycerol to produce desired products. Efforts have also been devoted to searching model compounds for glycerol in order to bridge the pressure and material gaps between model surfaces and powder catalysts. This work should provide insights into the rational design of more efficient catalysts for the upgrading of glycerol and potentially other biomass-derived polyols.

References

1 H. M. Mahmudul, F. Y. Hagos, R. Mamat, A. A. Adam, W. F. W. Ishak and R. Alenezi, *Renew. Sustain. Energy Rev.*, 2017, **72**, 497–509.

2 J. Hill, E. Nelson, D. Tilman, S. Polasky and D. Tiffany, *Proc. Natl. Acad. Sci.*, 2006, **103**, 11206–11210.

3 C. D'Agostino, G. Brett, G. Divitini, C. Ducati, G. J. Hutchings, M. D. Mantle and L. F. Gladden, *ACS Catal.*, 2017, **7**, 4235–4241.

530553 Oxophilic Metal Oxide Modified Iridium Catalysts for Selective Production of Renewable Hydrocarbons

Abstract:

Due to fossil fuel depletion, global warming and environmental problems, conversion of biomass into renewable hydrocarbons to supplement or gradually replace petroleum is highly desirable.[1] Hydrocarbons with different carbon range and structures (linear or branched) can be used as jet-fuel (C_8-C_{16}), diesel (C_9-C_{22}) and lubricant base oils (C_{20}-C_{50}). In recent years, great efforts have been devoted to the synthesis of various hydrocarbons from furan platforms by C-C coupling and hydrodeoxygenation (HDO). The HDO step is more challenging due to the involvement of high reaction temperature and pressure. We previously reported that rhenium oxide modified iridium (Ir-ReO_x/SiO_2) was an effective catalyst for the HDO of high carbon furylmethanes to C_{12}-C_{15} branched hydrocarbons.[2] However, rhenium is expensive and its replacement with a cheap metal can make renewable hydrocarbons cost-competitive and preserve rare metals. In this study, we show that a molybdenum oxide modified iridium catalyst (Ir-MoO_x/SiO_2) achieves comparable HDO performance as the Ir-ReO_x/SiO_2 and produces up to 90% jet-fuel range alkanes. Optimization of reaction conditions, catalyst recycling, extensive catalyst characterization and the reaction mechanism will be discussed.

1. W. Huber, S. Iborra, A. Corma, Chem. Rev., 2006, 106, 4044-4098.
2. S. Liu, S. Dutta, W. Zheng, N. S. Gould, Z. Cheng, B. Xu, B. Saha, D. G. Vlachos, ChemSusChem, 2017, 10, 3225-3234.

533616 Localizing Microwave Heat By Surface Polarization of Titanate Nanostructures for Enhanced Catalytic Reaction Efficiency

Abstract:

Titanate nanostructures (nanocube, nanotube and nanobelt) were successfully synthesized by using different alkalis in hydrothermal reaction. The titanate intermediates (protonated titanates, $H_2Ti_nO_{2n+1}$, n=3, 5) have been demonstrated critical in controlling the catalyst pore structure, surface area, crystal composition and the quantity of acid site. Especially, the open crystal structure of $H_2Ti_3O_7$ allows interlayer polarization of titanates, which is critical to enable a large number of Ti-O-SO_4^{2-} acid sites. The polarized titanate nanostructures can be used as microwave responsive catalysts for saccharide conversion reactions. The Ti-O-SO_4^{2-} not only serves as catalytic active site, but also the site for heat generation under microwave irradiation. Among the titanate nanostructures, titanate nanotube shows the best heat generation capability and gives the highest rate constant of 0.31 min^{-1}. The reaction equilibrium for HMF production can be reached within a few minutes at 140 °C. Benefited from the surface acidity and microwave heating ability, the energy efficiency of the reaction by titanate nanotube (5.6 mmol·$(KJ{\cdot}L)^{-1}$) is 9 times higher than commercial TiO_2 solid acid (0.6 mmol·$(KJ{\cdot}L)^{-1}$). The interlayer polarization is revealed as the major reason for the optimal microwave response and energy efficiency of the catalyst.

533968 Selective Glucose to Fructose Isomerization over Modified Zirconium UiOâ€66 in Alcohol Media

Abstract:

Biomass-derived feedstocks are alternative carbon sources to fossil fuels to produce value-added chemicals. Glucose isomerization to fructose is a key reaction for the conversion of sugars into these platform chemicals such as 5-hydroxymethylfurfural (HMF). Despite the use of zeolites as Lewis-acid catalysts for glucose isomerization, metal-organic frameworks (MOFs) have drawn attention due to their high tunability and density of active sites. In this work, we used a modified zirconium UiO-66, a highly stable MOF, as a catalyst for glucose isomerization reaction in alcohol media. The fructose selectivity is shown to change depending on solvent choice drastically. A combined effect of adsorption and solvation leads to the formation of alkyl-glucosides in depletion of fructose when methanol or ethanol are used. The use of 1-propanol as a solvent leads to selectivities of 72 % and 10 % in fructose and mannose, respectively at 82 % glucose conversion. We also demonstrate by ^{13}C NMR that the sugars adsorb onto UiO-66 in a closed form and that fructose is formed by an intramolecular C2-C1 hydride transfer mechanism.

[1] Dorneles de Mello, M.; Tsapatsis, M. ChemCatChem 10.1002/cctc.201800371.

535462 Production of Biorenewable Monomers - from Fructose to 2,5-Furandicarboxylic Acid

Abstract:

The demand for bio-based polymers is increasing globally due to their lower carbon footprint and independence from fossil fuels. 2,5-Furandicarboxylic acid (FDCA) is a potential bio-based alternative to terephthalic acid for the production of bio-based polymer polyethylene furanoate (PEF). However, the economical production of FDCA from carbohydrates must address technological challenges, such as instability of hydroxymethylfurfural (HMF), low solubility of FDCA, and requirement of stoichiometric amounts of base for the oxidation of HMF to FDCA.

We will report a process for production of FDCA from fructose which utilizes a solvent system comprising of a polar aprotic solvent and water to address the aforementioned challenges. We will discuss the role of solvent in stabilizing HMF, thereby leading to high yield. In our process, a high concentration of fructose (15 wt%) is dehydrated to HMF, and HMF is then oxidized to FDCA over a heterogeneous catalyst (Pt/C) without intermediate separation. We demonstrate that the Î³-Valerolactone (GVL)/H_2O solvent system has high FDCA solubility which not only leads to high FDCA yield but also eliminates the need of a homogeneous base during HMF oxidation. We further show that FDCA can be separated from the GVL/H_2O solvent system by crystallization, and a highly pure (>99% purity) FDCA product is obtained. Furthermore, we demonstrate that FDCA is an effective catalyst for fructose dehydration in GVL/H_2O, which obviates the need of a mineral acid catalyst, thereby improving both the economic and the environmental impact of the process. Our techno-economic model indicates that the overall process is economically competitive with current terephthalic acid processes.

537660 Comparative Analysis of Microalgae Productivity Potential and Economic Analysis in Open Raceway Ponds and Flat Panel Photobioreactors

Abstract:

Microalgae are often considered as promising feedstock for production of fuels and value-added chemicals. Microalgae can be cultivated in different kind of reactors. Open raceway ponds and flat panel photobioreactors are commonly used for microalgae cultivation in commercial scales. In the present work, a first-principles based quantitative model is developed to assess microalgae productivity potential in open raceways and flat panel photobioreactors for different locations in the United States. The mathematical growth model incorporates local climate factors and geospatial variabilities to determine microalgae productivity for the different reactors. Based on yearly microalgae productivity for the locations, a detailed economic analysis is conducted to assess microalgae price. A sensitivity analysis is further performed to assess the effect of different process parameters on microalgae price. Results indicate that there are significant differences in microalgae productivity and corresponding price for the different reactors and locations studied. The variation in algal productivity and costs can mainly be attributed to variation in local climate parameters and also to different engineering and process parameters pertaining to commercial scale microalgae cultivation. The quantitative modeling approach and corresponding economic analysis of microalgae cultivation will help to optimize the performance of microalgae-based systems at commercial scales for different geospatial locations.

524492 Subcritical Extraction of *Chlorella Pyrenoidosa*: Optimization through Response Surface Methodology

Abstract:

The objective of this study is to determine optimum operating conditions for maximum extraction of oil from subcritical methanol extraction of *Chlorella pyrenoidosa*. A statistical approach, i.e. response surface methodology (RSM) is employed in this study. The effects of three operational factors: reaction temperature (140 to 220 °C), reaction time (1 to 15 min) and methanol to algae ratio (1 to 9 wt.%) were investigated. A maximum oil yield of 7.1 wt.% was obtained at 160 °C, 3 min reaction time and 7 wt.% methanol to algae ratio. The analysis of variance (ANOVA) revealed that methanol to algae ratio is the most significant factor for maximizing oil yield. Regression analysis showed a good fit of the experimental data to the second order polynomial model.

529684 Lignin-Derived Deep Eutectic Solvents Pretreatment of Herbal Residues to EnhanceÂ Enzymatic Digestibility of Cellulose

Abstract:

A lignin-derived eutectic solvent was synthesized from choline chloride(ChCl) and p-coumaric acid(PCA). The results showed that after pretreatment with lignin-derived DES, the dense cellulose-hemicellulose-lignin structure of herb residue of *cortex albiziae*(HRCA) was effectively breakn down and this facilitated subsequent enzymatic hydrolysis. The optimum pretreatment condition was ChCl /PCA molar ratio of 1:1, 160°C-5 h, and solid-liquid ratio of 1:10. Compared with the untreated HRCA, 72h-enzymatic digestibility increased from 48.08% to 84.62%. When taking a water-DES pretreatment, the 72h digestibility level of the residue reached 99.36%. Moreover, it was found that the performance of DES pretreatment was closely related to the type of raw materials. Furthermore, the recycling and reusability of this novel DES were evaluated as well. Thus, synthesis and utilization of green, environment-friendly and recyclable lignin-derived DES would opened up a new way of lignin utilization and it has great potential in biomass refining.

535158 Evaluation of Structure, Topochemistry and Transport Reaction Processes in Plant Biomass during Pretreatment

Abstract:

Porous bio-based materials such as wood and paper have a complex internal porous structure that has a significant influence on the efficiency of further conversion processes, the properties of the materials synthesized and their end-use applications. Plant biomass is a renewable and sustainable raw material for a wide variety of bio-based products which include pulp and paper, bio-plastics, biofuels, wood plastic composites etc. The process of conversion of plant biomass to other useful products is heavily dependent on our understanding of the topochemical structure of plant cell walls and its influence on the reaction and transport properties. This is achieved through a combination of 2D and 3D characterization techniques such as Raman Spectroscopy, X-ray tomography (XRCT) and Transmission Electron Microscopy-Computed Tomography (TEM-CT).

The first step in the conversion process is the pretreatment step, which achieves the disruption of the recalcitrant structure of lignocellulose and increases the efficiency of the subsequent hydrolysis. One or more of the cell wall components are dissolved during this process which affects the structure, properties and topochemistry significantly. Hence is extremely important to evaluate these characteristics in order to develop effective pretreatment strategies. The topochemical changes in the plant cell walls during pretreatment are mapped using Raman spectroscopy. Based on this information, a transport reaction model based on a hybrid random walk and reaction is proposed to provide further insight into the pretreatment process. A fixed number of particles of the reagent used for the pretreatment process, which is representative of its concentration, diffuse through the lumen and pore spaces of the cell wall and follow a random walk path until they encounter the cell wall interface. At the interface, based on the probability of reaction and the ratio of diffusivities between the pore spaces and cell wall, they either react with one or more cell wall components or diffuse further. This stochastic dynamic approach keeps track of the spatial concentration of both the cell wall components and the reagent used for pretreatment in real time. This model can also be extended to 3D images obtained from TEM-CT. Additionally, changes in structural properties such as porosity, specific surface area and geometric tortuosity can be evaluated from TEM-CT images. Thus, using a combination of characterization techniques, it is possible to correlate the changes in 3D nanoscale architecture and topochemistry of the plant cell walls with the transport and reaction, thereby giving a fundamental insight into the pretreatment process. Ultimately, this would enable determination of effective conversion strategies not just for plant biomass, but for other similar porous materials as well.

Production of HMF from glucose using an Al^{3+} promoted acidic phenol-formaldehyde resin catalyst

Tingwei Zhang [a], Wenzhi Li [a,*]
[a] Department of Thermal Science and Energy Engineering, University of Science and Technology of China, Hefei 230026, PR China
*Corresponding author. Tel.: +86 0551 63600786
E-mail address: liwenzhi@ustc.edu.cn (Wenzhi Li)

Abstract: An Al^{3+} modified formaldehyde and p-hydroxybenzenesulfonic resin solid acid (Al-SPFR) was synthesized through a simple hydrothermal method, and was tested in the conversion of glucose into HMF. The morphology and texture structure of Al-SPFR was characterized by SEM, TEM and N_2 adsorption-desorption experiment, and the amount of acidic sites was measured by acid-base titration. Al-SPFR catalyst had higher specific surface area and strong acidic sites than that of previously prepared unmodified SPFR catalyst. The effects of reaction time and temperature, catalyst loading, and water content in water/GVL mixture were investigated for glucose dehydration. Over the Al-SPFR catalyst, an HMF yield of 47.4% was obtained at 170 °C for 2.0 h in the biphasic GVL/H_2O medium. Further performing a five-cycle experiment, the HMF yields initially decreased and then remained fairly constant, the recyclability of the Al-SPFR was reasonable.

1. Introduction

With the continuous endeavor for propelling sustainable development of human society and economy, numerous low-carbon technologies about wind, solar, renewable biomass, and nuclear energy have been exploited to adjust current heavily petroleum-dependant energy structure. Among these technologies, transforming lignocelluloses-derived carbohydrates into biofuels and biochemicals was considered as one of most promising supplements to our unsustainable fossil resources [1-4]. 5-hydroxymethylfurfural (HMF), a dehydrated product of C6 sugars, and because of its highly unsaturated structure, HMF has being emphasized as an important building block in the synthesis of a variety of liquid fuels, intermediate chemicals, and polymers, including 4-[5-(hydroxymethyl)furan-2-yl]but-3-en-2-one, levulinic acid, γ-valerolactone (GVL), 2,5-bis(hydroxymethyl)furan, 2,5-dimethylfuran, etc [5-7]. However, based on the present technical state-of-art, the large-scale production of HMF is still a difficulty, and it remains to be resolved.

Although significant yield of HMF has already been achieved in the dehydration of fructose over different kinds of heterogeneous or homogeneous acids, such as mineral acids, ion-exchange resin acids, carbon solid acids, and super solid acid, fructose content of natural biomass is relatively low with mainly storing in edible biomass, and commercialized fructose is commonly produced from glucose through the enzymatic isomerisation, thus fructose is not an ideal candidate for the production of HMF [3, 8-12]. Glucose is the most abundant C6 carbohydrate in nature, and directly converting glucose-based carbohydrates into HMF should be more economic [13].

Recently, laboratory-scale production of HMF from glucose is conducted by the cooperation of Lewis or basic and Brønsted acid catalysts, and such process involves two successive steps, the isomerisation of glucose into fructose, followed by the dehydration of fructose to HMF [13-16]. Homogeneous catalyst-containing systems include mineral acid/metal halide, mineral acid/solid Lewis acid, and metal halide /ionic liquids showed favorable activities for HMF production [17-19]. While the difficult of catalyst-solvent separation and catalyst recycle, and the high cost of the ionic liquids could lead the homogeneous catalyst not suitable in large-scale application. Therefore, aiming at improving the practicality, many heterogeneous acids include zeolites, functionalized silica and resins, modified heteropoly acids, metal phosphate, and super solid acids had been developed to prepare HMF from glucose in aqueous solution, organic solvents or biphasic water/organic solvent mixtures [20-25]. And some heterogeneous catalytic systems (zirconium phosphates in water, HMF yield of 46.6; silica–alumina composite THF/H_2O–NaCl, HMF yield of 63.1%) acquired as good HMF yield as those of Homogeneous systems. Among these various heterogeneous catalysts, functionalized resins, owning the features of easily mass preparation and relatively low cost, have great potential for industrialization.

So far, many functionalized polymer catalysts have been studied for the production of HMF. Tang et al. reported a sulfated cyclopentanone–formaldehyde condensate (SCFC) which was synthesized by one-step condensation/sulfation of formaldehyde, cyclopentanone, sodium sulfite and followed by cation-exchanging [26]. SCFC exhibited highly active for fructose dehydration, but producing HMF from glucose using SCFC was unsatisfactory, due to its absolute Brønsted acidity. A p-toluenesulfonic acid (PTSA) - paraformaldehyde (POM) resin acid PTSA-POM [27] previously was prepared by our group, and this catalyst showed similar performance to that of SCFS. HMF could be synthesized from glucose using solid Brønsted acid alone, though the yield of HMF obtained under this circumstance is relatively low, and lower than the results obtained using diluted mineral acids (HCl, H_2SO_4) in H_2O/GVL mixtures [28]. Cao et al. inspired by the isomerisation of glucose through phosphoglucose isomerase, designing an ammonium polymer P-BnNH_3Cl (P-BnNH_2 + HCl) for HMF production, the amino group in the catalyst promoted the ismerisation of glucose into fructose [29]. The highest HMF yield of 53% was reached in water:DMSO/methylisobutylketone (MIBK) mixtures at 140 ℃ in 10 h. Ravasco et al. modified Amberlyst-15 with $CrCl_3$, and the Cr loading onto Amberlyst-15 was 4.16 wt% [30]. The HMF yield achieved a maximum of 70% in tetraethyl ammonium bromide (TEAB)/water medium at 120 ℃ in 60 min. The resin-based catalysts have already developed extraordinary in the degradation of glucose. However, most prepared resin-based catalysts have small specific surface area (SSA) (PTSA-POM 2 m^2 g^{-1}; SCFC 4.6 m^2 g^{-1}). Solid acid catalysts with higher SSA may improve its efficiency to catalyze glucose conversion, increasing and achieving optimal HMF yield at shorter reaction time.

Another sulfonated phenol-formaldehyde resin catalyst (SPFR) prepared by the copolymerization of formaldehyde and p-hydroxybenzenesulfonic acid (PHSA) in acetic acid aqueous solution with the use of chitosan as soft template, had a SSA of

28.0 m^2 g^{-1} [31]. Using SPFR for HMF production from fructose and glucose in GVL/H_2O, HMF yield received from fructose was far prior to that of glucose. The small SSA of the SPFR should be one reason for the comparatively low yield of HMF from glucose, and if SPFR has higher SSA, and then its activities for HMF deriving from glucose may be better. Therefore, further attention should be put on the promotion of texture structure of SPFR. Zhang et al. reported applying $FeCl_3$ as the catalyst for the oxidative polymerization of triphenylamine to prepare aromatic polymer with good texture properties [25]. Therefore, we supposed that metal ions reacting with the component that involved in the synthesis of resins may improve resins' porous structure. Inspired by the above discussions, we supposed that introducing one metal ion that can chelate with chitosan or PHSA into the synthesis of SPFR may increase the SSA of SPFR, and thus improving its activity for producing HMF from glucose.

In this study, to avoid the possible pollution caused by the utilization of heavy metal ions of Fe, the Al ion was introduced into the process of SPFR preparation, though the Al^{3+} can merely chelate with chitosan. The Al-modified SPFR (named as Al-SPFR) was characterized by SEM, TEM, FT-IR, and N_2 adsorption-desorption experiment. The SSA increased more than twofold in Al-SPFR. And a fraction of Al immobilized in the prepared Al-SPFR may add some Lewis acidity, which is favorable towards the glucose isomerisation. Al-SPFR was tested for HMF production from glucose in GVL/H_2O, and the effects of the reaction time and temperature, catalyst loading, and water content in water/GVL mixture were investigated. Furthermore, the reuse of Al-SPFR was also examined.

2. Materials and methods

2.1 Materials

P-hydroxybenzenesulfonic acid hydrate (85%), D-(+)-glucose (GC, 99.5%), HMF (99%), sodium hydroxide standard solution (0.02 M), and chitosan (degree of deacetylation $\geq$ 95%) were supplied by Aladdin Industrial, Inc. (Shanghai, China). Acetic acid (GR), formaldehyde solution (AR, 37–40%), sodium chloride (NaCl, AR), $AlCl_3 \cdot 6H_2O$ (AR), and acetone (AR) were purchased from by Sinopharm Chemical Reagent Co., Ltd. GVL (95%) was purchased from LangFang Hawk Technology and Development Co., Ltd. All chemicals were directly used without further treatment.

2.2 Catalyst preparation

Al-SPFR was prepared via a modified hydrothermal method based on previously reported process by our group [31-32]. Typically, 0.45 g of chitosan and 0.5 g of $AlCl_3 \cdot 6H_2O$ were dispersed in 25 mL and 5 mL of distilled water (DIW) with magnetic stirring, respectively. And 0.6 mL of acetic acid was followed added into the chitosan suspension to obtain a pellucid solution. Afterwards, $AlCl_3$ solution was added slowly into the chitosan-containing acetic acid aqueous solution and stirred for half an hour. Then PHSA hydrate (3.1 g) was added to the mixture, after stirring for 10 min to completely dissolve PHSA, formaldehyde solution (3 mL) was lastly added and stirred for another 5 min. The mixture was enclosed in a 50 mL Teflon-lined autoclave, and placed into a preheated oven at 120 ℃ . After heating for 10 h, the autoclave was cooled to room temperature by means of natural cooling. The solid product was transferred to a beaker, and further oven-heated at 100 ℃ for 24 h. In the following washing process, the dried product was washed by firstly immersing in acetone/DIW under vigorous stirring, and then the product was washed by repeatedly immersing in DIW, until the filtrate was neutral. After washing, the product was oven-dried at 100 ℃ to constant weight.

2.3 Catalyst characterization

Structure information of the Al-SPFR was observed through SEM with energy dispersive X-ray (EDS) (FEI Sirion scanning electron microscope) and TEM (JEM 2010 electron microscope). N_2 sorption analysis was carried out in an Autosorb iQ (Quantachrome) apparatus. FT-IR spectra were recorded on a Thermo Scientific Nicolet iS-50 instrument (KBr disks). The amount of strong Brønsted acid sites of Al-SPFR was determined by acid-base titration [31]. Al-SPFR (0.1 g) was dispersed in saturated NaCl aqueous solution (20 mL) and stirred for 24 h to release the H^+ from Al-SPFR by cation-exchanging. The mixture was filtrated, and the residue was washed with saturated NaCl solution (3 mL*3) and distilled water (3 mL*3) in order. The washing liquor and filtrate were collected together, and the H^+ density in it was measured by base titration using sodium hydroxide standard solution (0.02 M).

2.4 Glucose conversion into HMF

The catalytic test was performed in 25 mL stainless steel reactor with magnetic stirring. In a typical synthesis of HMF, Al-SPFR (0.1 g), glucose (0.2 g), GVL (15.0 mL), and DIW (1.5 mL) were sealed in reactor. The reactor was heated to target temperature in 30 min, maintained at this temperature for a given time, and then the reactor was immerged into cooling water to timely terminate the reaction. The reaction solution was separated by filtration. The filtrate was gathered for HPLC detecting to obtain products yield.

2.5 Quantification procedure for HMF

Quantification analysis of products and unreacted glucose in samples were respectively performed by using HPLC (Shimadzu LC-10ADvp pump, equipped with a Shimadzu SPD-10Avp ultraviolet detector using an Agilent C18 column) and HPLC (Waters 515 pump, equipped with a Waters 2414 refractive index detector using a Bio-Rad HPX-87 column) [31, 33]. The contents of the products (furfural, HMF, and levulinic acid) and residual glucose were determined by standard calibration curves. Yields of products and conversion of glucose (mol%) were defined as:

Furfural yield = (moles of furfural produced / moles of starting glucose) × 100%

HMF yield = (moles of HMF produced / moles of starting glucose) × 100%

Levulinic acid yield = (moles of levulinic acid produced / moles of starting glucose) × 100%

Glucose conversion = (moles of reacted glucose / moles of starting glucose) × 100%

3. Results and discussion

3.1 Catalyst characterization

SEM and TEM images of the Al-SPFR are shown in **Fig.1**. Observing by SEM, Al-SPFR was consists of irregular particles with uneven micro surface, and the particle aggregation was somewhat serious, even so, compared with previously synthesized SPFR, Al-SPFR particles were smaller. TEM images also demonstrated the above conclusion, additionally, and it reveals that the particle aggregation of Al-SPFR particle was less severe than SPFR. The element detection of Al-SPFR was analyzed by EDS attached to the SEM (**Fig. 2**), which showed that Al-SPFR contained sulfur and aluminum elements. The S presented in Al-SPFR because of the PHSA, and indicating that S containing group remained stable during the hydrothermal process. The detectable Al suggested that a fraction amount of Al was remained in Al-SPFR by chelation with some residual chitosan. The SSA and total pore volume of the Al-SPFR is summarized in **Table1**. Surprisingly, both the SSA and Total pore volume of Al-SPFR were more than twice that of SPFR. This result was in good agreement with that of SEM and TEM, and this phenomenon could only be ascribed to the introduction of Al.

The FTIR spectrum of Al-SPFR is illustrated in **Fig. 3**. The broad band in the range of 3200-3670 cm^{-1} could be assigned to the O–H stretching and N–H stretching vibrations. Characteristic bands of methylene group appeared at 2926 cm^{-1} could also be received in Al-SPFR, which illustrated the successful synthesis of phenol-formaldehyde resin. The strong band centered at 1649 cm^{-1} could be ascribed to the amide I band, and the presence of N-containing group indicated that Al-SPFR contained a small amount of chitosan [31, 34]. Three characteristic bands of Al-SPFR distinctively at 1041 and 1274 cm^{-1} along with 1169 cm^{-1} could be attributed to the O=S=O symmetric stretching mode and SO_3–H stretching vibration, respectively, which are in accordance with the results of EDS. The amount of strong Brønsted acid sites of Al-SPFR was 0.96 mmol g^{-1}.

According to the previous studies, chitosan, dissolving in acetic acid aqueous solution to form transparent gels, were applied as a soft template for absorbing PHSA and being cross-linked with formaldehyde. The polymerization of PHSA and formaldehyde in acetic acid solution formed micromolecules and it was gradually deposited on chitosan gels, thus SPFR gels were finally generated. In this study, Al^{3+} was introduced into the above process, and then chitosan-Al complex was formed by reacting with the amino group and the hydroxyl group of chitosan. Thus, some active sites of chitosan, which should had originally reacted with formaldehyde and PHSA, are now occupied by Al^{3+}, Al^{3+} partially limited the chitosan's function as a template. And the chelation effect ensured Al^{3+} could be homodispersed in the SPFR gels, thus finally obtained particles had more uniform size. These should be the reason why the Al-SPFR particles were more uniform in size and the SSA is larger. Most of chitosan gels dissolved gradually in the hydrothermal process, but still a small amount of chitosan as well as chitosan-Al complex was remained in Al-SPFR gel [34].

3.2 Effects of reaction temperature and time of HMF production from glucose

Since Brønsted acidity of Al-SPFR takes up a significant part of total acidity, the temperature for glucose conversion over Al-SPFR was chosen in the relatively high range of 160-180 ℃ , and the reaction time in the range of 20-100 min was selected. As shown in **Fig.4A**, producing HMF from glucose was great influenced by temperature. When performing the reaction at 160 ℃ , the HMF yield increased during the whole selected reaction time. Upon increasing temperature to 170 and

180 ℃ , HMF yields increased firstly, after reaching optimal points, the HMF yields decreased gradually. And higher temperature accelerated the appearance of the optimal point. The maximum HMF yield of 41.5% was achieved at 180 ℃ in 50 min with glucose conversion of 96.2%. Apart from HMF, furfural and levulinic acid were detected in the reaction solution, either the yields of furfural or levulinic acid was increased with reaction temperature and time. Previous literatures reported that catalyzing by H_2SO_4, a large number of levulinic acid was synthesized in GVL/H_2O [35]. Therefore, it should be normal for the formation of some levulinic acid in our study. As for furfural, many researchers previously proposed that GVL/H_2O mixture was a special solvent for preparing furfural from glucose over acid catalyst [32, 36].

Although, under the optimal conditions, the yields sum of HMF, furfural and levulinic acid were more than 60%, and the Al-SPFR showed an improved activity for HMF production from glucose than using SPFR as the catalyst, but overall such HMF yield is not satisfactory. Glucose conversion increased with reaction temperature and time, and completely converted at 180 ℃ in 100 min. The selectivities toward HMF were relatively low, and many undesirable reactions should be synchronous occurred.

In the process of acid-catalyzed dehydration of glucose, hydroxyl groups of glucose were initially protonated to facilitate the formation of active intermediates with carbon cations and/or conjugated π bonds. These intermediates can finally convert to HMF or furfural, but can also go to the routine of polymerization. Furthermore, the rehydration of HMF would generate levulinic acid, and the intermediates formed during this process would be drawn into the polymerization [37-38]. Formation of soluble polymer and insoluble polymer by polymerization and cross-polymerization among glucose, furans (HMF and furfural), and reactive intermediates were complicated, further efforts should be made on clarifying it clear, which is conducive to achieving better HMF production from glucose.

Since getting as high HMF yield was set as unique goal, so other product including furfural and levulinic acid won't be referred in the rest of the experiments.

3.3 Effect of catalyst loading on HMF production from glucose

For the sake of disposing a certain amount of glucose using as little catalyst as possible and simultaneously maintaining preferable HMF yield, it is quite necessary to probe the effect of catalyst loading on HMF production from glucose. Fixing glucose usage at 0.2 g, Al-SPFR loading in the range of 0.05-0.4 g was applied, and the reactions were performed under the previously obtained optimal conditions (180 ℃ , 50 min). **Fig.5** shows the results, upon catalyst dose increasing from 0.05 g to 0.1 g, the HMF yields increased, and the optimal HMF yield was achieved at the Al-SPFR dose of 0.1 g. Upon further increasing catalyst dose to 0.4g, the HMF yields decreased during this period. The above indicated that neither too little nor too much of catalyst loading could provide a favorable HMF yield in a fixed reaction condition. The reduction in HMF yield at higher catalyst loading is due to a gain in side reactions, especially polymerization, caused by excessive acid sites. For example, a HMF yield of 37.6% was obtained using 0.2 g Al-SPFR at 180 ℃ in 50 min. While reacting at a milder condition (170 ℃ , 40 min), using the same glucose/Al-SPFR loading (0.2/0.2 g), given a better HMF yield of 43.4% (**Table 2**).

When fixing glucose usage at 0.2 g, the optimal Al-SPFR/glucose mass ratio was 0.5. However, upon reducing glucose usage to 0.1 g (Al-SPFR/glucose mass ratio of 1), the higher HMF yields were achieved at 170 ℃ (**Table 2**). These suggested that in a selected solvent volume, much more glucose loading may increase the occurrence opportunity of side reactions, and thus reducing the HMF yields.

3.4 Effect of water concentration in solvent on HMF production from glucose

Many previous literatures have proved that adding water into GVL could apparently improve the conversion of glucose into HMF, and the variation of water concentration in H_2O/GVL mixture significantly influence HMF production [13]. Therefore, the effect of blended solvents with different water contents (0%, 10%, 16.7%, 23.3%, and 100%, based on GVL) on producing HMF from glucose were investigated, and the results are summarized in **Table 3**. In terms of the effect water concentration in solvent on HMF production, Al-SPFR exhibited similar laws to those reported catalysts [13, 31]. Either pure water or pure GVL provided inferior HMFs yield to those of H_2O/GVL mixture; increasing water contents in H_2O/GVL mixture, the HMF yields decreased. H_2O/GVL volume ratio of 1/10 was the optimal value among the designated ratios.

GVL is a green and renewable solvent, and which has been identified as a promising solvent for the process of furans production. But present GVL production routine can't afford its large scale application. In this case, another industrial solvent, dimethylsulfoxide (DMSO) was also used for HMF production. Unfortunately, in comparison with H_2O/GVL, H_2O/DMSO didn't give comparable results under the same reaction conditions (**Table 2**).

3.5 Catalyst recycle

Catalyst recyclability is an important evaluation index for solid catalyst utilization, so reuse experiments of Al-SPFR for glucose dehydration were performed. In cycle 1, 0.15 g of Al-SPFR and 0.2 g glucose were added into the reactor, and reacted at 180 ℃ for 50 min. After the reaction, the spent catalyst was recovered by filtration, washing several times with acetone and oven-drying. Then the spent catalyst was used in the next cycle. As shown in **Fig.6**, the HMF yields decreased from cycle 1 to cycle 3, and remained fairly constant in the following cycle 3 to cycle 5. The FTIR spectrum of spent Al-SPFR revealed that the signal intensity of the peak centered at 1169 cm^{-1} (SO_3-H stretching vibration) became weaker than that of fresh catalyst, indicating that a detectable amount of SO_3-H was leached during the reuse experiments. Besides, a new characteristic peak appeared at 1711 cm^{-1}, which may be ascribed to the C=O stretching vibration. This phenomenon could be interpreted as the insoluble polymer byproducts with C=O group deposited onto the surface of Al-SPFR, and the insoluble polymers were produced by the polymerization reactions in which HMF participated [32, 39-40]. Both the loss of H^+ and the deposition of insoluble polymers onto the surface of Al-SPFR were the reasons for the decrease of HMF yields in recycle experiments. In general, the recyclability of the Al-SPFR was reasonable.

4. Conclusion

In summary, the introduction of aluminum ions into the synthetic process of p-hydroxybenzenesulfonic acid-formaldehyde resin acid catalyst is beneficial to improve the specific surface area and total pore volume of the catalyst, which given a more efficient catalytic conversion of glucose to HMF than the previous SPFR. The strong chelation of chitosan and Al^{3+} may be responsible for such an improvement, which partly limited the function of chitosan as a template. Using Al-SPFR as the catalyst, H_2O/GVL mixture provided higher HMF yield than in H_2O/DMSO. When the dose of Al-SPFR and glucose were both 0.1 g (Al-SPFR/glucose mass ratio of 1), an HMF yield of 47.4% was achieved at 170 ℃ in 120 min. The HMF yields initially decreased and then remained fairly constant during the reuse tests, the recyclability of the Al-SPFR was reasonable.

Acknowledgements

This study was financially supported by the Science and Technological Fund of Anhui Province for Outstanding Youth (1508085J01), the International Technology Cooperation Plan of Anhui (No.1503062030) and the National Key Technology R&D Program of China (NO. 2015BAD15B06).

References:

[1] Iris, K. M., Tsang, D. C. (2017). Conversion of biomass to hydroxymethylfurfural: A review of catalytic systems and underlying mechanisms. Bioresource Technol., 238, 716-732.

[2] Teong, S. P., Yi, G., Zhang, Y. (2014). Hydroxymethylfurfural production from bioresources: past, present and future. Green Chem., 16(4), 2015-2026.

[3] van Putten, R. J., Van Der Waal, J. C., De Jong, E. D., Rasrendra, C. B., Heeres, H. J., de Vries, J. G. (2013). Hydroxymethylfurfural, a versatile platform chemical made from renewable resources. Chem. Rev., 113(3), 1499-1597.

[4] Mukherjee, A., Dumont, M. J., Raghavan, V. (2015). Sustainable production of hydroxymethylfurfural and levulinic acid: Challenges and opportunities. Biomass and Bioenergy, 72, 143-183.

[5] Shi, N., Liu, Q., Zhang, Q., Wang, T., & Ma, L. (2013). High yield production of 5-hydroxymethylfurfural from cellulose by high concentration of sulfates in biphasic system. Green Chem., 15(7), 1967-1974.

[6] Lin, H., Xiong, Q., Zhao, Y., Chen, J., Wang, S. (2017). Conversion of carbohydrates into 5 - hydroxymethylfurfural in a green reaction system of CO2 - water - isopropanol. AIChE Journal, 63(1), 257-265.

[7] Zhang, Y., Pan, J., Chen, Y., Shi, W., Yan, Y., & Yu, L. (2016). HIPEs template: towards the synthesis of polymeric catalysts with adjustable porous structure, acid–base strength and wettability for biomass energy conversation. Chem. Eng. J., 283, 956-970.

[8] Zhao, J., Zhou, C., He, C., Dai, Y., Jia, X., & Yang, Y. (2016). Efficient dehydration of fructose to 5-hydroxymethylfurfural over sulfonated carbon sphere solid acid catalysts. Catal. Today, 264, 123-130.

[9] Li, Y., Liu, H., Song, C., Gu, X., Li, H., Zhu, W., Yin, S., Han, C. (2013). The dehydration of fructose to 5-hydroxymethylfurfural efficiently catalyzed by acidic

ion-exchange resin in ionic liquid. Bioresource Technol., 133, 347-353.
[10] Choudhary, V., Mushrif, S. H., Ho, C., Anderko, A., Nikolakis, V., Marinkovic, N. S., Frenkel, A. I., Sandler, S. I., Vlachos, D. G. (2013). Insights into the interplay of Lewis and Brønsted acid catalysts in glucose and fructose conversion to 5-(hydroxymethyl) furfural and levulinic acid in aqueous media. J. Am. Chem. Soc., 135(10), 3997-4006.
[11] Román-Leshkov, Y., Chheda, J. N., Dumesic, J. A. (2006). Phase modifiers promote efficient production of hydroxymethylfurfural from fructose. Science, 312(5782), 1933-1937.
[12] Graça, I., Bacariza, M. C., & Chadwick, D. (2018). Glucose isomerisation into fructose over Mg-impregnated Na-zeolites: Influence of zeolite structure. Micropor. Mesopor. Mat., 255, 130-139.
[13] Zhang, T., Fan, W., Li, W., Xu, Z., Xin, H., Su, M., Lu, Y., Ma, L. (2017). One - Pot Conversion of Carbohydrates into 5 - (Hydroxymethyl) furfural using Heterogeneous Lewis - Acid and Brønsted - Acid Catalysts. Energy Technol., 5(5), 747-755.
[14] Nikolla, E., Román-Leshkov, Y., Moliner, M., & Davis, M. E. (2011). "One-pot" synthesis of 5-(hydroxymethyl) furfural from carbohydrates using tin-beta zeolite. Acs Catal., 1(4), 408-410.
[15] Zhang, Y., Xiong, Q., Chen, Y., Liu, M., Jin, P., Yan, Y., Pan, J. (2018). Synthesis of Ceria and Sulfated Zirconia Catalysts Supported on Mesoporous SBA-15 toward Glucose Conversion to 5-Hydroxymethylfurfural in a Green Isopropanol-Mediated System. Ind. Eng. Chem. Res., 57(6), 1968-1979.
[16] Bermejo-Deval, R., Assary, R. S., Nikolla, E., Moliner, M., Román-Leshkov, Y., Hwang, S. J., Palsdottir, A., Silverman, D., Lobo, F., Curtiss, L., Davis, M. E. (2012). Metalloenzyme-like catalyzed isomerizations of sugars by Lewis acid zeolites. Proc. Natl. Acad. of Sci., 109(25), 9727-9732.
[17] Gallo, J. M. R., Alonso, D. M., Mellmer, M. A., Dumesic, J. A. (2013). Production and upgrading of 5-hydroxymethylfurfural using heterogeneous catalysts and biomass-derived solvents. Green Chem., 15(1), 85-90.
[18] Bali, S., Tofanelli, M. A., Ernst, R. D., Eyring, E. M. (2012). Chromium (III) catalysts in ionic liquids for the conversion of glucose to 5-(hydroxymethyl) furfural (HMF): insight into metal catalyst: ionic liquid mediated conversion of cellulosic biomass to biofuels and chemicals. Biomass and bioenergy, 42, 224-227.
[19] Heguaburu, V., Franco, J., Reina, L., Tabarez, C., Moyna, G., & Moyna, P. (2012). Dehydration of carbohydrates to 2-furaldehydes in ionic liquids by catalysis with ion exchange resins. Catal. Commun., 27, 88-91.
[20] Huang, F., Su, Y., Tao, Y., Sun, W., Wang, W. (2018). Preparation of 5-hydroxymethylfurfural from glucose catalyzed by silica-supported phosphotungstic acid heterogeneous catalyst. Fuel, 226, 417-422.
[21] Chen, D., Liang, F., Feng, D., Xian, M., Zhang, H., Liu, H., Du, F. (2016). An efficient route from reproducible glucose to 5-hydroxymethylfurfural catalyzed by porous coordination polymer heterogeneous catalysts. Chem. Eng. J., 300, 177-184.
[22] Li, X., Peng, K., Xia, Q., Liu, X., & Wang, Y. (2018). Efficient conversion of

cellulose into 5-hydroxymethylfurfural over niobia/carbon composites. Chem. Eng. J., 332, 528-536.
[23] Saravanan, K., Park, K. S., Jeon, S., Bae, J. W. (2018). Aqueous Phase Synthesis of 5-Hydroxymethylfurfural from Glucose over Large Pore Mesoporous Zirconium Phosphates: Effect of Calcination Temperature. ACS Omega, 3(1), 808-820.
[24] Li, X., Xia, Q., Peng, K., Liu, X., Essayem, N., Wang, Y. (2016). High yield production of HMF from carbohydrates over silica–alumina composite catalysts. Catal. Sci. Technol., 6(20), 7586-7596.
[25] Zhang, L., Xi, G., Zhang, J., Yu, H., Wang, X. (2017). Efficient catalytic system for the direct transformation of lignocellulosic biomass to furfural and 5-hydroxymethylfurfural. Bioresource Technol., 224, 656-661.
[26] Tang, H., Li, N., Chen, F., Li, G., Wang, A., Cong, Y., ... & Zhang, T. (2017). Highly efficient synthesis of 5-hydroxymethylfurfural with carbohydrates over renewable cyclopentanone-based acidic resin. Green Chem., 19(8), 1855-1860.
[27] Li, W., Xu, Z., Zhang, T., Li, G., Jameel, H., Chang, H. M., Ma, L. (2016). Catalytic conversion of biomass-derived carbohydrates into 5-hydroxymethylfurfural using a strong solid acid catalyst in aqueous γ-valerolactone. BioResources, 11(3), 5839-5853.
[28] Li, M., Li, W., Lu, Y., Jameel, H., Chang, H. M., Ma, L. (2017). High conversion of glucose to 5-hydroxymethylfurfural using hydrochloric acid as a catalyst and sodium chloride as a promoter in a water/γ-valerolactone system. RSC Adv., 7(24), 14330-14336.
[29] Cao, X., Teong, S. P., Wu, D., Yi, G., Su, H., Zhang, Y. (2015). An enzyme mimic ammonium polymer as a single catalyst for glucose dehydration to 5-hydroxymethylfurfural. Green Chem., 17(4), 2348-2352.
[30] Ravasco, J. M., Coelho, J. A., Simeonov, S. P., Afonso, C. A. (2017). Bifunctional Cr^{3+} modified ion exchange resins as efficient reusable catalysts for the production and isolation of 5-hydroxymethylfurfural from glucose. RSC Adv., 7(13), 7555-7559.
[31] Li, W., Zhang, T., Xin, H., Su, M., Ma, L., Jameel, H., Chang, H., Pei, G. (2017). p-Hydroxybenzenesulfonic acid–formaldehyde solid acid resin for the conversion of fructose and glucose to 5-hydroxymethylfurfural. RSC Adv., 7(44), 27682-27688.
[32] Zhang, T., Li, W., An, S., Huang, F., Li, X., Liu, J., Pei, G., Liu, Q. (2018). Efficient transformation of corn stover to furfural using p-Hydroxybenzenesulfonic acid-formaldehyde resin solid acid. Bioresource Technol. 264, 261-267.
[33] Li, W., Zhang, T., Pei, G. (2018). Catalytic Conversion of Corn Stover into Furfural over Carbon-based Solid Acids. BioResources, 13(1), 1425-1440.
[34] Yu, Z. L., Wu, Z. Y., Xin, S., Qiao, C., Yu, Z. Y., Cong, H. P., Yu, S. H. (2014). General and straightforward synthetic route to phenolic resin gels templated by chitosan networks. Chem. Mat., 26(24), 6915-6918.
[35] Alonso, D. M., Wettstein, S. G., Mellmer, M. A., Gurbuz, E. I., & Dumesic, J. A. (2013). Integrated conversion of hemicellulose and cellulose from lignocellulosic biomass. Energy Environ. Sci., 6(1), 76-80.
[36] Gürbüz, E. I., Gallo, J. M. R., Alonso, D. M., Wettstein, S. G., Lim, W. Y., Dumesic, J. A. (2013). Conversion of hemicellulose into furfural using solid acid

catalysts in γ - valerolactone. Angew. Chem. Int. ed., 52(4), 1270-1274.
[37] Hu, X., Kadarwati, S., Wang, S., Song, Y., Hasan, M. M., Li, C. Z. (2015). Biomass-derived sugars and furans: which polymerize more during their hydrolysis?. Fuel Process. Technol., 137, 212-219.
[38] Tsilomelekis, G., Orella, M. J., Lin, Z., Cheng, Z., Zheng, W., Nikolakis, V., Vlachos, D. G. (2016). Molecular structure, morphology and growth mechanisms and rates of 5-hydroxymethyl furfural (HMF) derived humins. Green Chem., 18(7), 1983-1993.
[39] van Zandvoort, I., Wang, Y., Rasrendra, C. B., van Eck, E. R., Bruijnincx, P. C., Heeres, H. J., Weckhuysen, B. M. (2013). Formation, molecular structure, and morphology of humins in biomass conversion: influence of feedstock and processing conditions. ChemSusChem, 6(9), 1745-1758.
[40] Reiche, S., Kowalew, N., Schlögl, R. (2015). Influence of synthesis pH and oxidative strength of the catalyzing acid on the morphology and chemical structure of hydrothermal carbon. ChemPhysChem, 16(3), 579-587.

Figures and Tables

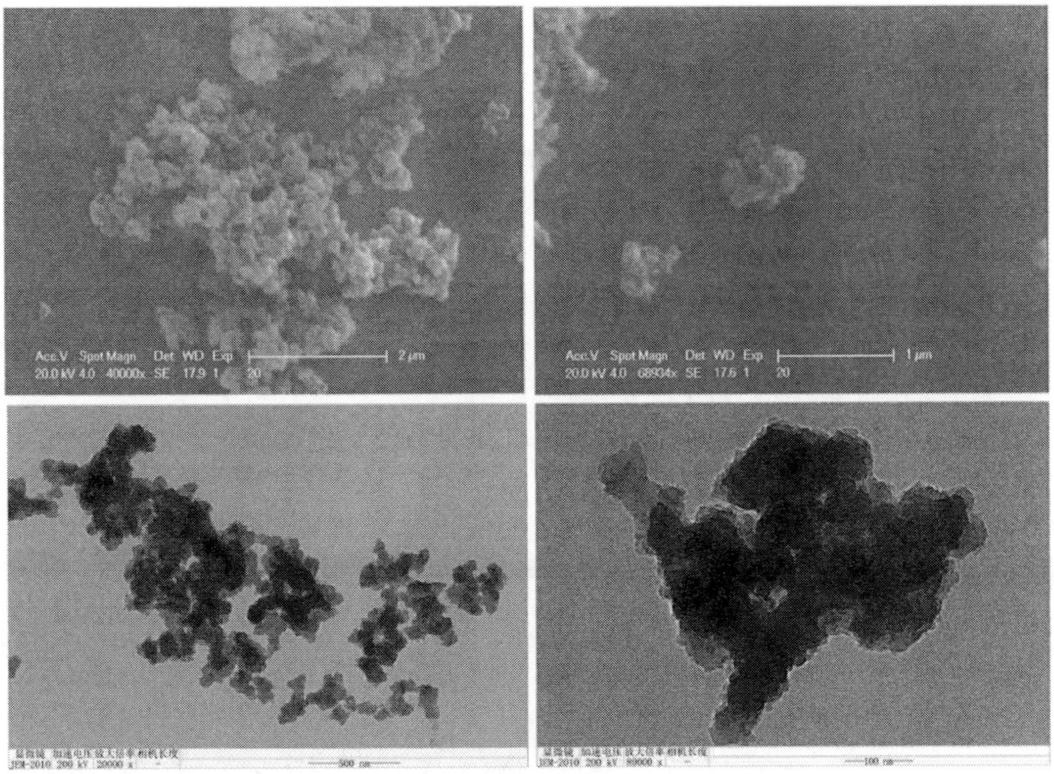

Fig.1 SEM and TEM images of the Al-SPFR.

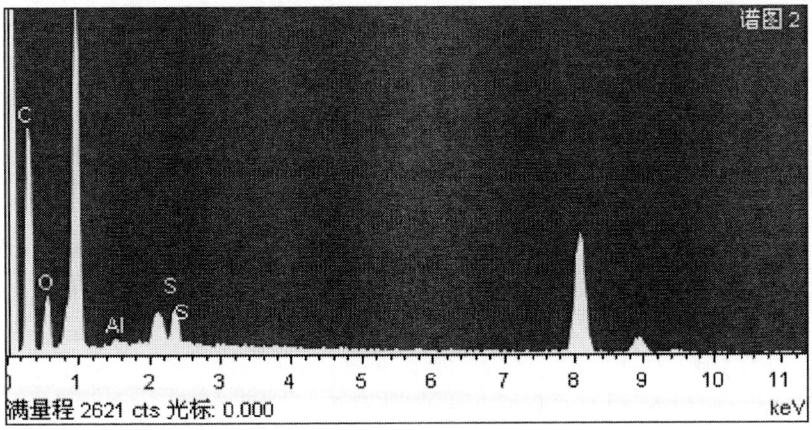

Fig.2 Energy dispersive X-ray spectrum of the Al-SPFR.

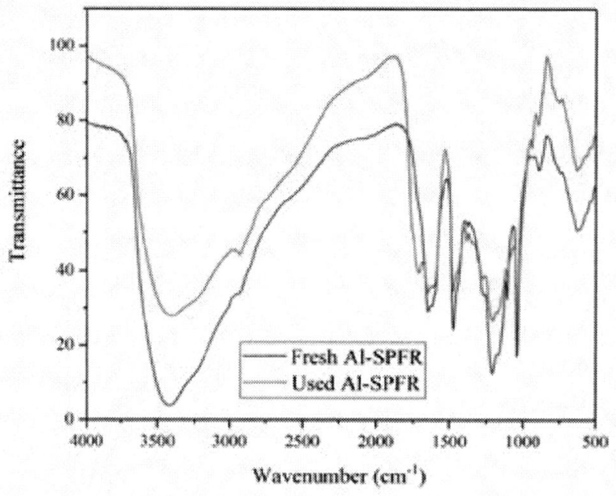

Fig.3 FTIR spectra of the fresh Al-SPFR and used Al-SPFR (five times).

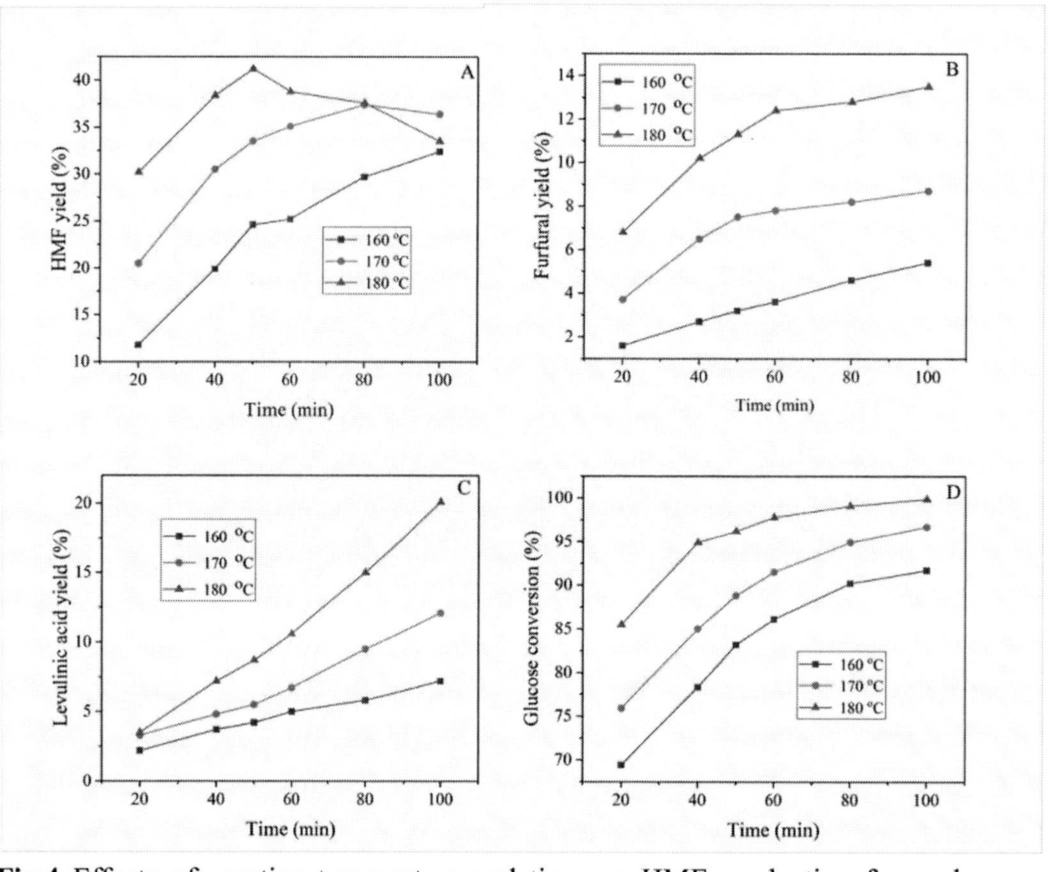

Fig.4 Effects of reaction temperature and time on HMF production from glucose. Reaction conditions: 0.20 g of glucose, 0.1 g of Al-SPFR, GVL/DIW (15/1.5 mL).

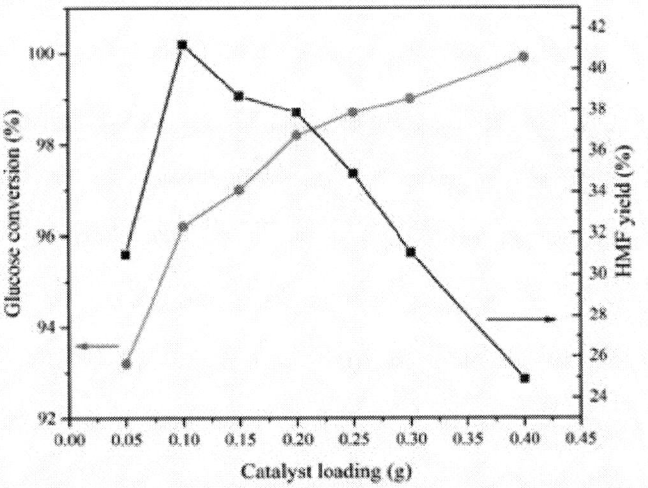

Fig.5 Influence of catalyst loading on glucose conversion, and HMF yields. Reaction conditions: glucose (0.2 g), GVL/DIW (15/1.5 mL), 180 °C, 50 min.

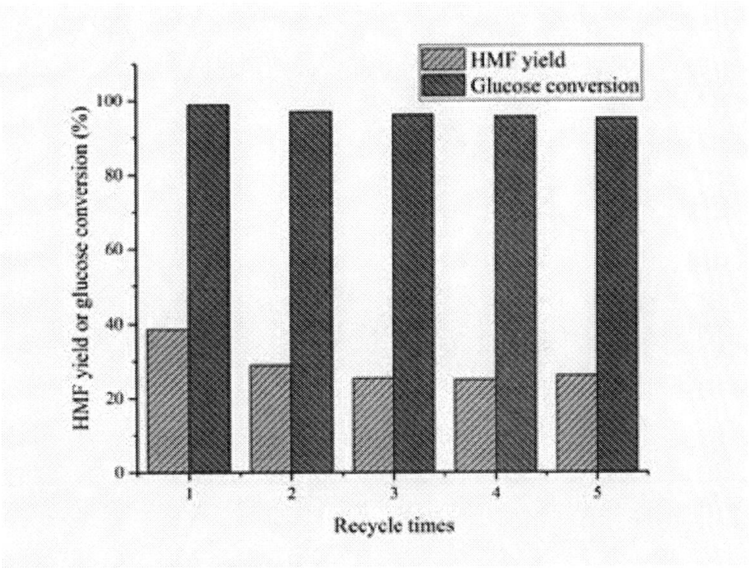

Fig.6 Cycle usage of the Al-SPFR catalyst for HMF production from glucose. Reaction conditions: 0.20 g of glucose, 0.15 g of Al-SPFR, GVL/DIW (15/1.5 mL), 180 ℃ , 50 min reaction time.

Table 1. Structural parameters of Al-SPFR and previously prepared SPFR.

Parameters	Al-SPFR	SPFR-0.38	MSPFR
Specific surface area (m^2/g)	71.5	27.98	24.8
Total pore volume (cm^3/g)	0.20	0.11	0.08
Refs.	This study	31	32

Table 2. HMF yield from different Al/glucose mass ratio or using different solvents.

Entry	Organic solvent/H_2O v/v (mL)	Temperature and time	Catalyst/glucose loading (g)	HMF yields (%)
1	GVL/DIW 15/1.5	170 °C, 40 min	0.2/0.2	43.4
2	GVL/DIW 15/1.5	170 °C, 60 min	0.1/0.1	42.1
3	GVL/DIW 15/1.5	170 °C, 120 min	0.1/0.1	47.4
4	DMSO/DIW 15/1.5	170 °C, 40 min	0.1/0.2	15.3
5	DMSO/DIW 15/1.5	170 °C, 60 min	0.1/0.2	17.8
6	DMSO/DIW 15/1.5	170 °C, 120 min	0.1/0.2	32.6

Table 3. Influence water content in solvent on HMF yield from glucose [a].

Entry	Water addition (mL)	Temperature and time	HMF yields (%)	Glucose conversion (%)
1	Pure GVL [b]	180 °C, 50 min	30.8	99.0
2	1.5	180 °C, 20 min	30.2	85.5
3	1.5	180 °C, 50 min	41.2	96.2
4	1.5	180 °C, 80 min	37.6	99.1
5	2.5	180 °C, 20 min	28.7	82.9
6	2.5	180 °C, 50 min	36.8	94.4
7	2.5	180 °C, 80 min	33.0	98.5
8	3.5	180 °C, 20 min	23.6	81.7
9	3.5	180 °C, 50 min	33.6	91.2
10	3.5	180 °C, 80 min	31.7	97.1
11	Pure water [c]	180 °C, 50 min	17.7	69.9

[a] Reaction conditions: 0.1 g Al-SPFR, 0.2 g glucose, GVL/DIW (15/1.5 mL). [b] None of DIW was added into the reactor. [c] None of GVL was added into the reactor.

546232 Low Temperature and High Efficiency Biomass Fuel Cell and Bio-Hydrogen Production

Abstract:

A novel fuel cell which can directly use native polymeric biomasses, such as starch, cellulose, lignin, and even switchgrass and wood powders will be discussed. This fuel cell combines some features of solar cells, fuel cells, and redox flow batteries. Specifically, an oxidation/reduction ion pair, such as Fe^{2+}/Fe^{3+}, or polyoxomatelate, are used as catalyst to oxidize biomass by either absorbing solar light or heat energy. The power density of the solar-induced hybrid fuel cell powered with raw biomass could reach ~50 mW/cm^2. Unlike most cell technologies that are sensitive to impurities, the cell reported in this study is inert to most organic and inorganic contaminants present in the fuels. The fuel cell is completely noble metal free. The similar fuel cell system can also be converted to electrolysis cell for hydrogen production at low temperature using native biomass directly. Either polyoxometalate (POM) or $FeCl_3$ was used as the catalysts and charge transfer agents in anode. Over 90% Faraday efficiency was achieved. In a thermal insulation reactor, the heat energy can be maintained at a very low stage for continuous operation. Compared to the best alkaline water electrolysis reported in literature, the electrical energy consumption could be 40% lower with novel lignin electrolysis method. Our results suggest that POM or $FeCl_3$ mediated electrolysis process can significantly reduce the electrolysis energy consumption in hydrogen production and, simultaneously, depolymerize lignin to low molecular weight value-added aromatic chemicals.

533650 Nano Core-Shell Structured ZSM-5@Mesoporous Silica for Catalytic Co-Cracking Phenolic Bio-Oil Model Compounds and Ethanol to Aromatics

Abstract:

As the depletion of fossil fuels and severe pollution caused by the excessive usage become an upcoming threat, looking for alternative and green energy sources is an urgent matter for sustainable development. Lignocellulosic biomass can be converted to fuels and aromatics as industrial chemicals and the essential additive in jet fuel through feasible methods, such as fast pyrolysis. However, because of its high content of phenolic compounds, the upgrading of this kind of bio-oil remains as a big challenge. Co-cracking raw bio-oil with alcohols is considered to be an effective method. But the conventional catalyst ZSM-5 shows an unsatisfied catalytic performance and still needs to be improved. The fast deactivation of catalysts caused by severe coke formation on the catalyst particlesâ€™ surface and the low selectivity of desired aromatics product are the biggest issues.

The nano-sized molecular sieve catalysts have showed a better performance in many areas because of its bigger surface area and improved diffusion of reactants and products. Also the core-shell structured ZSM-5 with mesoporous shell has proved its potential in adsorption, hydrocracking, and supercritical catalytic cracking and so on. However, when combining the two techniques together to rebuild ZSM-5, more studies need to be done to reveal the influence of the catalyst properties on catalytic performance. Here, the effect of mesoporous shell and crystal size on the conversion of bio-oil and selectivity over aromatics products is studied by preparing nano ZSM-5 with different size by seed-induced method and core-shell composites by a simple hydrothermal treatment in alkaline solution with cetyltrimethylammonium bromide (CTAB), noted as ZSM-5@meso SiO_2. The size of synthesized composites are about 350nm, 700nm and 1Î¼m, respectively. And the thickness of mesoporous silica shell is about 25nm. Six model compounds were used to represent the phenolic pyrolysis bio-oil, and then mixed with ethanol. The mixture had 50wt% of ethanol, 25wt% of phenol, guaiacol and eugenol, and 25wt% of acetic acid, hydroxyacetone and methylfuran. The conventional ZSM-5(Si/Al=50) had only 4.47% (GC-MS peak area %) of BTX products and 83.27% of phenolic products. While the nano ZSM-5 had much higher selectivity on aromatic and BTX, up tp 75% higher than those of ZSM-5. And eugenol can be completely converted. By tailored with the mesoporous shell, the ZSM-5@meso SiO_2 can effectively suppress the selectivity of PAH. The ratio of MAH to PAH for core-shell structure was about 20% higher than those of catalysts not tailored. Under a typical reaction condition(450â„ƒ, WHSV=$2h^{-1}$), the nano core-shell ZSM-5@meso SiO_2 could have a selectivity of 24.39% over aromatics and only 55.3% over phenolic products, with guaiacol and eugenol completely converted. The improved catalytic performance of nano core-shell composites can offer a new route for designing novel cracking catalysts with better resistance to coke formation and higher hydrocarbon selectivity.

533775 Does Recrystallization in Aqueous Environment Affect the Reactivity of Ball-Milled Cellulose for Acid Catalyzed Hydrolysis?

Abstract:

Lignocellulosic biomass is a potential renewable feedstock that could at least partially displace fossil resources for the production of industrially valuable fuels and chemicals. A biorefinery scheme for conversion of lignocellulosic biomass consists of preatreatment followed by enzymatic hydrolysis. However, the main hurdle to economic utilization of lignocellulose is cost-effective depolymerization of cellulose, one of the major components containing the majority of carbohydrates, to glucose. Thus, significant research effort has been spent to gain fundamental understanding of cellulose reactivity and develop and design processes to improve its conversion.

Cellulose is a homopolymer of glucose units connected by Î²-1,4 glycosidic bonds; polymer chains form dense hydrogen bond network with their hydroxyl groups, resulting in a highly crystalline solid resistant to dissolution and chemical conversion. Typical depolymerisation to glucose is through acid or enzyme catalysed hydrolysis. Kinetic analyses have attempted to relate cellulose hydrolysis rate to structural parameters such as crystallinity, degree of polymerization, particle size and surface area. Literature reports suggest that cellulose crystallinity is the major factor determining reactivity. Models explain this hypothesis by differential rate of depolymerisation of amorphous and crystalline regions. Thus, a large number of pretreatment processes aiming at cellulose decrystallization have been proposed to decrease time and condition severity of cellulose conversion. However, an overlooked phenomenon is cellulose recrystallization in aqueous environment and most hydrolysis processes happen in water solutions. In this work, we studied the relationship between recrystallization of ball-milled cellulose and its difficulty to depolymerize in acid catalysed hydrolysis.

We used vibratory ball-milling of cellulose for varying periods of time to generate samples with different degrees of crystallinity. Hydrolysis with hydrochloric acid confirmed that progressive ball-milling resulted in higher rate of depolymerisation to glucose, consistent with the decrease in crystallinity. According to the current model, recrystallization of cellulose should result in a decrease of hydrolysis rate. We used hydrothermal treatment of ball-milled samples to increase the crystallinity, avoiding cellulose hydrolysis at the treatment conditions. Further, we hydrolysed the recrystallized cellulose and compared their glucose yield to ball-milled only samples. However, the glucose yield showed only slight decrease, not consistent with the increase in crystallinity.

The lack of full recalcitrance recovery suggested that there is another structural feature of ball-milled cellulose that was determining its reactivity. Particle analysis revealed fast initial decrease of size followed by a stabilization during ball-milling. Thus, it could not explain increase in hydrolysis rate of progressively ball-milled samples. Further structural analysis with gel permeation chromatography and electron spin resonance revealed that ball-milling results in chain breaking and generation of new chain ends. We hypothesize that these are accessible and highly active sites for acid catalysed hydrolysis.

Our results suggest that crystallinity is not always the major parameter affecting cellulose hydrolysis rates and depolymerisation models should be updated to include recrystallization and other structural features. Furthermore, the obtained fundamental knowledge motivates new cellulose pre-treatment processes and use of hydrolysis conditions that do not result in recrystallization.

535007 Solvent and Processing Conditions for Pretreatment and Dissolution of Cotton Cellulose

Abstract:

Efficient utilization of biomass is hindered by the recalcitrance to dissolution of semicrystalline cellulose. Pretreatment is often used to alter the structure of cellulosic biomass in order to make cellulose more accessible to solvents and enzymes. The pretreatment involves physical and/or chemical processing which affects the degree of crystallinity and size of biomass particles. We examine here the effects of (i) solvent properties, pretreatment steps and temperature, and (ii) fiber diameter and degree of crystallinity, on the kinetics of cellulose swelling and dissolution. To this end we have combined (a) experimental results on cotton fiber swelling, change in crystallinity and dissolved amount when treated under different solvent conditions, with (b) a phenomenological model that accounts for the phenomena governing the dissolution of solid cellulose, e.g., solvent penetration, transformation from crystalline to amorphous domains, specimen swelling, and polymer chain untangling. [Ghasemi, M.; Singapati, A. Y.; Tsianou, M.; Alexandridis, P., Dissolution of semicrystalline polymer fibers: Numerical modeling and parametric analysis. AIChE Journal 2017, 63 (4), 1368-1383. DOI: 10.1002/aic.15615] The insights obtained from this analysis would facilitate the rational selection of solvents and the design of pretreatment processes to reduce the size and degree of crystallinity of cellulosic biomass particles, leading to enhanced biomass utilization. [Ghasemi, M.; Tsianou, M.; Alexandridis, P., Assessment of solvents for cellulose dissolution. Bioresource Technol. 2017, 228, 330-338. DOI: 10.1016/j.biortech.2016.12.049] [Ghasemi, M.; Alexandridis, P.; Tsianou, M., Dissolution of cellulosic fibers: effect of crystallinity and fiber diameter. Biomacromolecules 2018, 19 (2), 640-651. DOI: 10.1021/acs.biomac.7b01745]

537994 Synthesis of Artificial Lignin Polymers and Their Effects on Enzymatic Hydrolysis of Cellulose

Abstract:

Lignin is the second abundant polymer present in lignocellulosic biomass. Dimeric lignin model compounds have been often used to investigate lignin degradation of lignin and catalytic conversion. However, these models are difficult to represent the three dimensional complex structure of lignin polymers and their effects on enzymatic hydrolysis of cellulose.

The lignin model polymers were synthesized to contain β-O-4 and β–β linkages that are structural motifs ubiquitous in native lignin obtained from hardwood. The method mainly consists of 4 parts: 1) preparation of monomeric acetate, 2) multiple-step synthesis of β–β dimeric diacetate, 3) construction of β-hydroxyl ester-containing oligomers via lithium diisopropylamide (LDA)-promoted aldol type reaction of β–β dimeric diacetate with monomeric acetate, and 4) reduction of the ester moiety with $LiAlH_4$ to give the lignin model polymer. The combined yield of the last two steps is 27%. The β–β linkage is from β–β dimeric diacetate while the β-O-4 linkage is the repeating unit of this polymer.

The synthesized lignin polymers were characterized by gel permeation chromatography and HSQC NMR. The number average degree of polymerization (DP_n) of the polymer is 8.25 and the molecular weight of lignin polymers were 2378. These artificial lignin polymers decreased the 72 h hydrolysis yield of Avicel from 57% to 53%, which is similar to the effect of native lignin on Avicel.

548227 Effect of lignin alkylation in organosolv pretreatment on enzymatic hydrolysis

Abstract:

Understanding the interactions between lignin and cellulase enzymes will help better design biomass pretreatment and enzymatic hydrolysis processes. Ethanol organosolv lignins (EOLs) were isolated from organosolv pretreatment of sweetgum (SG) and loblolly pine (LP) under different temperature. The structural changes of EOLs were determined by ^{13}C NMR and HMBC NMR to elucidate their positive and/or negative effects on enzymatic hydrolysis. Alkylation of lignin with ethanol in organosolv pretreatment resulting in the etherification of Cî± was observed in both EOL-SG and EOL-LP. The effect of EOL-SG on enzymatic hydrolysis was correlated with the content of ethoxyl group. The lignin alkylation in EOLs appeared to be reversible at high temperature, resulting in the inverse change of their effects on enzymatic hydrolysis. A strong correlation was observed between the binding affinity of cellulases onto EOLs and their corresponding hydrolysis yields. Lignin alkylation likely reduce the affinity of enzyme onto lignin, which can enhance the enzymatic hydrolysis.

525592 Effects of Warm Water Washing on the Fast Pyrolysis of *Arundo Donax*

Abstract:

Effects of Warm Water Washing on the Fast Pyrolysis of *Arundo Donax*

D.S. Chandler, F.L.P. Resende

Arundo donax, also known as Giant Reed, is a fast growing herbaceous reed native to India, but pervasive throughout the tropics and subtropics. It grows faster than bamboo, miscanthus and switchgrass at a rate of 30 dry tons/acre/year [1]. The fresh water requirements of *A. donax* are much lower than those of many other potential bio-energy crops. Moreover, fast pyrolysis of an herbaceous weed such as *A. donax* can mitigate problems such as agricultural displacement associated with the use of other forms of biomass [2].

A major hurdle in fast pyrolysis of herbaceous biomass is its high inorganic content, specifically the very high content of alkali salts. The problems associated with pyrolysis of materials that contain inorganics are cracking of volatiles, char formation, and catalyst poisoning [3–7]. Many of these inorganic compounds act as unintended catalysts that crack the volatile pyrolysis products into smaller compounds, mainly permanent gases like CO_2, CO, and CH_4. They also promote the formation of char, though it is unclear if this takes place by increasing the rates of condensation reactions or by reducing the heat transfer rate to the particles [4,5,8].

In this talk, we will report how warm water washing of biomass affects fast pyrolysis. Specifically, we applied this concept to *Arundo donax*, a common herbaceous biomass, and studied the washing effects on fast pyrolysis yields in a Py-GC/MS and a fluidized bed reactor. We stir-washed samples in water at 20°C, 40°C and 60°C, and found that water at 60°C led to an 18% increase in the removal of ash and an 11% increase in the removal of potassium compared to a room temperature wash. We pyrolyzed these samples via Py-GC/MS-FID and found that the reduction in ash increased the pyrolysis yields of acetol, furanone and levoglucosan relative to the unwashed samples. Additionally, we pyrolyzed washed *A. donax* in a fluidized bed reactor and compared the results to those of unwashed *A. donax*. We found that *A. donax* washed at 60°C produced 9.7 wt% more bio-oil than unwashed, with increases in glycolaldehyde and levoglucosan yields of 0.78 wt% and 1.24 wt%, respectively. The yield of olefinic gases increased by 0.22 wt% and yields of carbon dioxide and alkane gases decreased 2.40 wt% and 0.11 wt% respectively. Char and water yields decreased 4.32 wt% and 10.08 wt% respectively. The gas produced from pyrolysis of washed *A. donax* has a greater HHV than that from unwashed *A. donax* (7.77 MJ/kg vs. 6.81 MJ/kg). We believe these results are a consequence of differences in the extent of secondary reactions catalyzed by the inorganic species present in herbaceous biomass.

[1] Tree Free Biomass Solutions, Growing NileFiber, (n.d.). http://www.treefreebiomasssolutions.com/growing-nile-fiber (accessed January 1, 2016).

[2] C. Williams, T. Biswas, A. Downie, Yield , salt tolerance and energy from Arundo donax : a potential biochar and biofuel crop, in: 1st Asia Pacific Biochar Conf., Queensland, AU, 2009.

[3] B. Gudka, J.M. Jones, A.R. Lea-Langton, A. Williams, A. Saddawi, A review of the mitigation of deposition and emission problems during biomass combustion through washing pre-treatment, J. Energy Inst. 89 (2016) 159–171. doi:10.1016/j.joei.2015.02.007.

[4] P.R. Patwardhan, R.C. Brown, B.H. Shanks, Understanding the product distribution from biomass fast pyrolysis, Iowa State University, 2009. doi:10.1016/j.jaap.2009.08.007.

[5] L. Deng, T. Zhang, D. Che, Effect of water washing on fuel properties, pyrolysis and combustion characteristics, and ash fusibility of biomass, Fuel Process. Technol. 106 (2013) 712–720. doi:10.1016/j.fuproc.2012.10.006.

[6] A. Trendewicz, R. Evans, A. Dutta, R. Sykes, D. Carpenter, R. Braun, Evaluating the effect of potassium on cellulose pyrolysis reaction kinetics, Biomass and Bioenergy. 74 (2015) 15–25. doi:10.1016/j.biombioe.2015.01.001.

[7] F.A. Agblevor, S. Besler, Inorganic compounds in biomass feedstocks. 1. Effect on the quality of fast pyrolysis oils, Energy & Fuels. 10 (1996) 293–298. doi:10.1021/ef950202u.

[8] J. Huygye, F. Ronsse, W. Prins, The role of alkali and earth alkaline metals as intrinsic catalysts in the fast pyrolysis of hemicellulose, 2015. http://lib.ugent.be/fulltxt/RUG01/002/217/034/RUG01-002217034_2015_0001_AC.pdf.

533648 Ex-Situ Catalytic Cracking of Biomass Pyrolysis Vapors over Montmorillonite K10-Supported Iron (III) Oxide

Abstract:

Bio-oil produced from biomass pyrolysis exhibit low stability, corrosiveness, high viscosity, and low calorific value. Pyrolysis vapor upgrading can improve the stability and other properties of bio-oils to facilitate storage, transport, and drop in use as fuels and chemical feedstocks in existing petroleum refining infrastructure. Synthetic acid zeolites, in particular HZSM-5, have shown good catalytic cracking results for pyrolysis vapors, however, synthesis cost and catalyst deactivation due to susceptibility to coking limit its application for pyrolysis upgrading. A naturally occurring catalyst material with mesoporous structure could reduce costs and prolong catalyst use by reducing deactivation by coking.

In this study, catalytic cracking of biomass pyrolysis vapor over montmorillonite K10 (MMT) supported iron catalyst was investigated in a fixed bed lab scale reactor. Iron loadings of 0%, 5%, and 10% Fe (weight basis) were prepared by wet impregnation and calcined catalysts were characterized by several methods. Fast pyrolysis was carried out in an inductively heated reactor at 600 °C under 1.25 L min^{-1} nitrogen flow, and the vapors were passed over a heated catalyst bed (360 °C, 1.5 catalyst-to-biomass ratio) prior to condensation by a cold trap and electrostatic precipitator. Non-catalytic pyrolysis was also carried out for comparison.

Char yield remained constant for all runs (18.1+/-0.9) as pyrolysis conditions were kept the same. Compared to non-catalytic pyrolysis, catalytic upgrading resulted in lesser liquid yield and greater non-condensable gas yield, due to catalytic cracking. Liquid yields from upgrading were found to be 17.8, 14.0, and 17.5% for pyrolysis over 0, 5, and 10% Fe/MMT, respectively, compared to 53.6% liquid yield for non-catalytic pyrolysis. Water yield in the bio-oil increased with increasing Fe content, indicating promotion of dehydration reactions over catalysts with greater Fe. Non-condensable gas yields were 54.3, 57.0, and 51.3% for 0, 5, and 10% Fe/MMT, respectively, while non-catalytic pyrolysis gas yields were 27.5%.

As a result of upgrading, yields of polyaromatics in the bio-oil were found to decrease indicating occurrence of cracking reactions over the catalysts studied. Upgraded bio-oil is composed of mainly phenols, aromatic and aliphatic ketones, and furans/pyrans. Non-upgraded bio-oil contains phenols as well, but also contains less stable molecules such as carboxylic acids and aldehydes. On average, the non-condensable gas produced from catalytic pyrolysis is composed of 15.2% CH_4, 51.9% CO, 18.5% CO_2, 4.7% H_2, and 5.7% C2-C6. According to the carbon balance, we saw 50% carbon conversion to the gas fraction as a result of upgrading compared to 30% conversion for non-catalytic pyrolysis. Energy balance revealed greater gas higher heating values of 160 MJ/kg (dry biomass basis) compared to that of non-catalytic pyrolysis gas (90 MJ/kg, dry biomass basis). Carbon conversion to coke decreased from 1.4% on pure montmorillonite K10 catalyst to 0.4% for 10% Fe catalyst.

536729 Iron-Based Chemical Looping Biomass Gasification for Carbon Nanofiber Production: Process Simulation and Experiments

Abstract:

Carbon nanofibers (CNFs) are widely used in multiple industrial areas such as electrochemistry, catalysis, material manufacture, etc. Nowadays, researches are intensively focused on developing cost-effective and environmentally friendly pathways to produce high-quality CNFs. The Ohio State University (OSU) has been collaborating with Applied Science Inc. (ASI) to investigate a chemical looping based pathway for low-cost production of CNFs as a replacement for high grade carbon black. The pathway under investigation consists of two consecutive steps: chemical looping biomass to syngas (BTS) process developed by OSU and syngas to CNFs conversion process developed by ASI. The synergistic combination of the two processes enables the conversion of a clean and renewable raw material, woody biomass, into a value-added product via an energy- and cost-efficient pathway, substituting the use of natural gas as the feedstock in conventional processes. This study includes simulation and experimental works. ASPEN Plus is utilized to simulate the OSU BTS process with respect to the operating conditions including temperature, oxygen carrier/fuel feed ratio and H_2O, CO_2 feed rates and their influence on the syngas composition. Syngas with various H_2/CO ratio and CO/CO_2 ratio based on simulation results will then be injected into CNF reactors at ASI to synthesize CNFs and compare the product performance under different operating conditions including syngas feed composition, temperature and the use of microwave. It is shown from simulation that OSU BTS process is able to produce a flexible H_2/CO ratio ranging from 0.4 to 2. Preliminary experiments conducted at ASI under various temperature and feedstock inputs have shown a mass yield of 19% using CO as the main feedstock, higher than the 16% yield achieved using CH_4 as the feedstock in conventional processes. Additionally, CNFs produced from CO shows a ~10 times increase in specific volume compared to CNFs produced from natural gas, CH_4, suggesting that the CNFs produced from CO could be smaller in diameter and better-ordered. Further experiments will be conducted using CO, CO2 and H_2 at different concentration as the feedstock to investigate their impact on CNF yield and quality. In conclusion, the combination of OSU BTS and ASI CNF production process shows a significant improvement over the traditional technologies by lowering cost, enhancing carbon conversion efficiency, reducing CO_2 emissions and has significant market penetration potential.

530195 Techno-Economic Analysis of Simultaneous Biomass Gasification and Syngas Upgrading Via Chemical Looping Technology

Abstract:

This study is to analyze a chemical looping gasification system for the production of high-quality syngas from biomass. The system includes two reactors: one for biomass gasification and the other one for catalytic syngas cleaning and oxygen carrier regeneration operated as a two-stage process. One of the innovations of this research is to develop and use an iron-based material in a redox cycle to transfer oxygen and heat for biomass gasification and to serve as a catalyst for hot-syngas cleaning. Another innovation is to co-produce high-quality biochar from biomass gasification itself by recycling CO_2 and H_2O in syngas as a fluidizing and activating agent in the redox cycle. We also recycle CO_2 directly into the gasifier and explore the ratio of steam to CO_2 as the co-gasifying agent.

We are developing an Aspen Plus process model for techno-economic analysis of the production of Fischer-Tropsch liquids via biomass gasification using this novel chemical looping technology. The process models will be validated by experimental data and used to investigate the energy efficiency, productivity, and energy and material flows of the integrated chemical looping biomass gasification system with various process configurations and operations for producing syngas at various qualities from various biomass feedstocks. An economic model will be developed and incorporated into the process model to determine the cost of the chemical looping system at different production scales. The capital costs will be determined by the components of the system, their basic costs and the life of the system. The operating costs will be determined by thermal energy and electricity consumption as well as labor and maintenance costs.

529616 Study of Effect of Reaction Conditions on the Hydrothermal Liquefaction Reaction Followed By Steam Reforming of the Liquefaction Liquid

Abstract:

Concerns are raised due to the excessive use of fossil fuels. The drastic depletion of the natural resources has urged a need to find an alternative for energy production. Hydrothermal liquefaction (HTL) is found to be the suitable process available to tackle the wet feed stock conversion to fuel energy. Reduction of the drying steps reduces the feed preparation cost which gives HTL a substantial advantage over other thermochemical routes. The use of water as a solvent for breakdown of the biomass in the presence of high temperatures is termed as hydrothermal liquefaction. High moisture, sub and supercritical water conditions enables the process to breakdown the structure and yield gas, aqueous and solid phase. This abstract highlight the effect of reaction temperature and feed composition on the seaweed conversion and yield of various phases. The effect of reaction temperature was observed at a constant seaweed: water ratio of 1:10 with an increase of 40°C from the starting temperatures of 180°C and a fixed reaction time of 2 hours. The reaction was performed at 4 different operating temperature conditions i.e., 180°C, 220°C, 260°C and 300°C. Heating rate varied during the experiment from 1.3, 0.7, 1.1 and 1.1 °C/min respectively. Maximum conversion of seaweed, 91.7 % was obtained at a temperature of 300°C. The effect of seaweed to water ratio was later observed at 300°C. The ratio was varied, and experiment was run at 4 different seaweed: water at 1:5, 1:7 and 1:10 at a reaction temperature of 300°C with a constant reaction time of 2 hours and a varying heating rate of 0.8, 0.8 and 1.1°C/min respectively. The highest conversion of 91% was observed at seaweed: water of 1:10 at a temperature of 300°C. Steam Reforming of the liquefaction liquid obtained at a seaweed: water of 1:10 at a reaction temperature of 300 °C for 2 hours was carried out a temperature range of 600 °C to 800 °C. Effect of reaction temperature on hydrogen yield and carbon gases conversion was observed at a temperature of 600 °C, 700 °C and 800 °C. Comparison for the product gas composition was observed at the reaction temperatures. It was observed during the experiments that with increasing feed ratio, the liquid color was changed from dark brown to brown. Reaction temperature had more effect on the conversion rather than the feed ratio on the liquefaction reaction. Hydrogen yield was increased with an increase in the temperature and was observed to be highest at a temperature of 800 °C.

Acknowledgment: This research was respectfully supported by Engineering Development Research Center (EDRC) funded by the Ministry of Trade, Industry & Energy (MOTIE). (No. N0000990)

534187 Effect of Distinctive Detoxification on Inhibitors Removal and Butanol Fermentation of Poplar Prehydrolysates

Abstract:

The prehydrolysates from dilute acid biomass pretreatment contains fermentable sugars for biofuels, however, carbonyl degradation compounds present severe toxicity to the fermentation microbes. Furans (furfural and hydroxymethylfurfural), aliphatic acids (acetic acid, formic acid and levulinic acid) and phenolic compounds have been suggested to be the main inhibitors in biomass prehydrolysates. However, the most potent inhibitors have not been identified and the distinctive effect of various detoxification methods on a broad spectrum of inhibitory compounds has been less studied.

Effect of distinctive overliming and activated carbon (AC) detoxification on inhibitors removal and butanol fermentation of the poplar prehydrolysates was investigated. Gas chromatography-mass spectrometry (GC/MS) was used to analyze the degradation compounds, and 46 carbonyl compounds were identified as potential inhibitors including 2,5-furandicarboxyaldehyde and 2,5-hexanedione. It was observed that overliming and AC treatment alone could not make the prehydrolysates fermentable with *Clostridium*. The sequential overliming and AC resulted in remarkable fermentability and high butanol yield at 0.22 g g$^{\hat{a}€``1}$ sugar. Synergistic effect of overliming and AC detoxification on ABE fermentation was observed. We also examined the compositional changes in the prehydrolysates treated by overliming and AC. Overliming removed 75.6% of furan derivatives and 68.1% of aromatic monomers. In comparison, AC (5.0% *w/v*) removed 77.9% of furan derivatives and 98.6% of aromatic monomers. In addition, overliming removed much more 2,5-furandicarboxyaldehyde, 5-ethylfuran-2-carbaldehyde and 2,5-hexanedione than AC did. On the contrary, AC could remove considerably more phenolic acids than overliming. Our results showed cinnamic acid, dihydroxybenzoidc acid, coumaric acid, vanillic acid and ferulic acid could be removed by 100%, 100%, 100%, 85.74% and 10%, respectively after AC treatment. While they were reduced only by 74.69%, 13.99%, 1.13%, 32.61% and 60.94%, respectively after overliming. In the sequential detoxification, both dialdehydes/diketones and phenolic acids were extensively removed. This probably was the main reason why the sequential detoxification enabled a remarkable ABE fermentation for the prehydrolysates. Aliphatic acids (formic, acetic and levulinic acids) was not changed in the sequential detoxification.

This study revealed that the distinctive effect of overliming and AC treatment on the prehydrolysates detoxification. Overliming removed more dialdehydes and diketones than AC treatment, while AC removed more phenolic acids than overliming. Sequential overliming and AC treatment were required to make the prehydrolysates fermentable with *Clostridium saccharobutylicum*. The study also suggested different detoxification method was needed for ABE fermentation of the prehydrolysate as comparing ethanol fermentation.

536748 Increased Lactic Acid Production in Fermentation Process from Woody Biomass and Its Kinetic Modelling

Abstract:

Currently, lactic acid has been widely utilized in food industry, chemical industry and poly (lactic acid) synthesis. However, low cost lactic acid is required for the successful commercial development of cost competitive lactate esters as substitutes to those derived from petrochemical sources. Hemicellulose from woody biomass, with its renewability and readily availability, offers one of the most economically favorable substrate alternatives for biotechnological production of optically pure lactic acid. In this work, a structured kinetic model of batch fermentation of lactic acid from woody biomass is developed and the effects of operational conditions on productivity are studied, which can be further optimized for increased lactic acid production.

529999 Carbon-Based Solid Acid Pretreatment in Corncob Saccharification: Specific Xylose Production and Efficient Enzymatic Hydrolysis

Abstract:

In this paper, we disclose a novel saccharification technology for lignocellulosic biomass. A new carbon-based solid ($C-SO_3H$) acid catalyst was first synthesized by a simple, one-step hydrothermal carbonization method using microcrystalline cellulose and sulfuric acid. The functional group, chemical composition and structure of the catalyst were characterized. After five reuses, the solid acid catalyst still showed a high catalytic activity for corncob pretreatment. Under optimal pretreatment conditions (140 °C, 6 h, 0.25g corncob, 0.25g catalyst and 25ml water), xylose was directly released from corncob in a high yield (78.1%). Enzymatic hydrolysis of the pretreatment residue provided an enzymatic digestibility of up to 91.6% in 48 h. The structure, morphology, and components of the corncob and residues were analyzed. The high xylose and glucose yields confirmed the high catalytic activity of the synthetic carbon-based solid acid, providing green and effective lignocellulose utilization.

536383 Hydrolysis of Cellulose and Its Adsorption Performance on Heavy Metal Irons

Abstract:

With the rapid development of the economic society and industry, water pollution of heavy metal irons had become one of the most serious environmental problems facing the world. Conventional treatment technologies for the removal of these toxic heavy metals are not economical and further generate huge quantity of toxic chemical sludge. In recent years, cellulose due to its abundance, low cost and ease of operation, are increasingly used in heavy metal iron removal.

In my research, the effect of hydrolysis on the adsorption properties of cellulose was studied. We optimize the reaction conditions of cellulose hydrolysis and study the kinetics of the adsorption process. Hydrolysis of cellulose is further studied to improve the adsorption performance.

527555 Introduction of Self-Steam Explosion Pretreatment Technology for Wet Biomass

Abstract:

The main obstacles to the widespread use of lignocellulosic biomass include its high moisture content and low grindability. Thus, we herein report a novel method, viz. the self-steam explosion (SE) method, which utilizes the moisture content of biomass to both reduce the particle size and enhance it properties. More specifically, the moisture content of the biomass sample is fully utilized as a steam resource to auto-hydrolyze the biomass component and yield fine particles with diameters < 1 mm through self-explosion. This process differs from conventional steam explosion, which mainly employs saturated steam from external resources and requires waste water treatment due to contamination of the steam, as the self-SE method offers flexibility and recyclability in terms of the heating resources required in addition to a significant reduction in wastewater production. In this study, we demonstrated this self-SE method effectively reduced the particle sizes of hardwood, softwood measuring 16â€"22 mm to produce fine particles. The moisture content and chips size of biomass affect the milling performance of this method. Self-SE method produces higher yield of fine particles and consumes less energy than conventional steam explosion method.

527232 Non-Equilibrium Plasma Pretreatment of Biomass for Enhanced Conversion

Abstract:

Non-equilibrium plasma pretreatment of biomass for enhanced conversion

Lusi A, Haiyang Hu, Hui Hu, Xianglan Bai

Plasma treatment has been emerged as a green, low-cost alternative for biomass pretreatment. Compared to conventional biomass pretreatment, plasma treatment does not require aggressive solvents or chemicals. Instead, plasma discharge is generated by simply applying electricity. Non-equilibrium plasma is particularly attractive due to its low energy input and mild condition. The ions, electrons and radicals generated by the plasma discharge can create chemically rich environment to promote various reactions in biomass, such as delignification, depolymerization, hydrolysis and oxidation. The effect of plasma discharge also highly depends on the gas environment and plasma conditions. In the present study, the effects of plasma pretreatment on biomass pyrolysis were investigated. More specifically, biomass and biomass compositions were subjected to either AC plasma or nanosecond pulsed plasma using DBD reactor under atmospheric air or argon. The effect of moisture was also investigated by adding water during the plasma pretreatment. The plasma pretreated biomass and biomass components were subsequently pyrolyzed, and the product distributions were compared with the results from pyrolysis of untreated biomass. In the present study, pyrolysis of plasma pretreated cellulose produced an unseen high levoglucosan yield up to 73.4%.

537083 *Effect of Aromatic Additives on Dilute Acid Pretreatment of Aspen*

Abstract:

Biomass pretreatment is essential for producing biofuels from lignocellulosic biomass. The aromatic additives (such as *p*-hydroxybenzoic acid and 2-napthol) have the potential to prevent lignin repolymerization in pretreatment and improve enzymatic hydrolysis. Effects of *p*-hydroxybenzoic acid and 2-naphthol additives on dilute acid pretreatment were investigated. In this study, we examined the enzymatic hydrolysis of the pretreated substrates and the physicochemical properties and structures of residual lignin. It was observed that *p*-hydroxybenzoic acid and 2-naphthol improved the 72h hydrolysis yield of dilute acid pretreated substrates by 13% and 17%, respectively. Langmuir adsorption isotherms showed that *p*-hydroxybenzoic acid and 2-naphthol decreased the residual lignin affinity to cellulase enzymes and reduced the distribution coefficient from 0.232 L/g to 0.124 and 0.084 L/g, respectively. The residual lignin molecular weight was determined by gel permeation chromatography. The potential reaction between lignin and *p*-hydroxybenzoic acid and 2-naphthol was characterized by HSQC-NMR. The study indicated that aromatic additives could change lignin hydrophobicity and reduce lignin condensation in biomass pretreatment and enhance the enzymatic hydrolysis.

531806 Fractionation, Conversion, and Valorization of Lignocellulosic Biomass in Inorganic Ionic Liquid (molten salt hydrate)

Abstract:

Molten salt hydrate, also called inorganic ionic liquid, is an aqueous solution of an inorganic salt at extraordinarily high concentration. Molten salt hydrate has many unique properties such as high boiling point, low vapor pressure, and liquid state at moderate temperature. Different from organic ionic liquids, molten salt hydrates have low cost, low viscosity, and low toxicity. More attractively, some molten salt hydrates are able to swell and dissolve cellulose and biomass. Therefore, they can be used as a solvent and reaction medium for biomass conversion and valorization. This presentation reports our recent research activities on using molten salt hydrates for fractionation, conversion, and valorization of lignocellulosic biomass, including cellulose dissolution and hydrolysis, biomass saccharification and fractionation, glucose isomerization, sugars/biomass to furan-based chemicals and hydrocarbons, lignin quantitation, and lignin depolymerization.

532109 Are Lignocellulosic Feedstocks Commercially Relevant to Make Pure Sugars for Chemical Catalytic Upgrading?

Abstract:

Significant research has been done for upgrading underutilized cellulose, hemicellulose, and lignin of lignocellulosic feedstock using chemocatalytic technologies to fuels and chemicals. One requirement for chemocatalytic upgrading of monomeric/polymeric sugars derived from cellulose and hemicellulose is that these sugars must be nearly pure. However, the currently proposed technologies for the biomass fractionation/pretreatment were developed with a focus of making unpurified sugars for the biocatalytic upgrading. One aim of this work is to select potential biomass pretreatment/fractionation technologies to make pure sugars for chemocatalytic upgrading through two-level screening. Following the two-level screening, the combined autohydrolysis and organosolv (AO) and formic acid pulping technologies were found to be suitable for the effective fractionation of lignocellulosic feedstock to cellulose, hemicellulose, and lignin. The optimal plant scales for the fractionation of lignocellulosic feedstock using the AO and formic acid pulping technologies were determined. The AO and formic acid pulping technologies were simulated using ASPEN Plus to determine the energy consumption for the biomass fractionation technologies. The minimum selling price (MSP) of making purified sugars from a lignocellulosic feedstock was estimated for both AO and formic acid pulping technologies following the discounted cash flow methodology. The estimated MSP of sugars produced via AO and formic acid pulping technologies are \$ 0.5 per kg and \$ 0.8 per kg, respectively. The requirement of a high liquid to solid ratio is found to be a limiting factor for attaining the target sugar price of \$ 0.3 per kg with both AO and formic acid pulping technologies.

526555 Catalytic Etherification of Glycerol to Glycerol Oligomers in the Presence of Alumina Supported Ca/Sr Mixed Oxides

Abstract:

In this work, the solvent-less etherification of glycerol to oligomers using alumina supported Ca and Sr catalysts was studied. These catalysts were obtained with impregnation of strontium and calcium onto γ-alumina (γ-Al_2O_3) in $Sr(NO_3)_2$ and $Ca(NO_3)_2$ solutions. The evidence for synergistic effects between active Ca and Sr species was demonstrated by etherification tests over a series of Ca-Sr-Al catalysts with varied Ca-to-Sr molar ratios. Effect of calcination temperatures of these supported catalysts, from 600 °C to 900 °C, on the resultant reusability and conversion efficiency of glycerol to oligomers and the selectivity of linear and branched oligomer to cyclic ones were investigated. The catalyst properties were characterized by various instruments including BET, TPD, XRD and SEM, and the ICP-OES analysis was performed to confirm the leaching of the metal species in the liquid phase of the reaction mixture. With Ca and Sr supported on γ-alumina catalyst, the catalyzed etherification reactions were carried out at 250 °C with 2.5 wt% Ca/Sr/γ-alumina calcined at 900 °C in nitrogen atmosphere. Under optimal impregnation conditions, a conversion efficiency of glycerol 80 % glycerol could be obtained after 10-h reaction. Moreover, a selectivity near 70% of di- and tri-glycerol in linear and branched forms over the cyclic form could be achieved.

538721 The Role of Copper Stability in Selectively Condensation of Ethanol to Higher Alcohols

Abstract:

The recent emergence of a strong bioethanol industry has elevated interest towards its use as a sustainable platform molecule towards producing high valued chemicals; one such process is the selective conversion of ethanol into high value C_{4+} alcohols. This multistep reaction involving acid-base catalyzed condensation and hydrogen transfer tends to yield a complex mixture of side products such ethers, esters, and ketones that prevent efficient conversion. Our work has demonstrated 80% selectivity to higher alcohols with conversion above on MgAl hydrotalcite catalysts at low copper loadings, with stable catalyst lifetimes over 100 hours in a flow through-reactor setup. Lower concentrations of copper was shown to prevent the formation of ester and ketone side products without reducing the conversion of ethanol. Characterization of the catalyst by STEM and XAS EXAFS work reveal that this is largely due to high dispersion of copper at lower copper loadings with near absence of copper clustering, even under high temperatures and reductive conditions. Further *in operando* XANES and EXAFS study shows that copper only reduces partially to Cu

support matrix allow for the production of dispersed Cu^{+1} sites, which promotes the efficient catalytic reaction to higher alcohols while not contributing to side reaction pathways such as esterification and C-C scission. The dispersion and stability of these individual copper promoters is critical to the high efficiency of this reaction. This work will discuss the limiting copper movement within the support to lengthen the lifetime of catalyst activity as well as throughout multiple regeneration cycles.

538786 Effects of Alloying Pd and Cu on Tandem Dehydrogenation-Aldol Condensation Reactions

Abstract:

The successful reduction of carbon dioxide emissions as a result of the production and consumption of fuels and chemicals requires the use of biomass resources as a feedstock. As a result of this, the discovery of novel pathways to upgrade platform molecules to fuel precursors is of critical importance. More specifically, for the production of diesel substitutes, the formation of new C-C bonds is required, as platform molecules deriving from the acid-catalyzed hydrolysis of biomass or the fermentation thereof are in the C2-C6 carbon range.

In this work, we investigate the upgrade of fermentation-derived acetone-butanol-ethanol mixtures to renewable diesel precursors. To achieve this, we develop a bimetallic Pd-Cu catalyst. We show that alloying of Pd and Cu results in the suppression of undesirable decarbonylation reactions. We present X-ray absorption spectroscopy results showing the segregation of Cu to the catalyst nanoparticle surface. Kinetic measurements under flow conditions demonstrate that Pd-Cu surface ensembles are active for hydrogenation reactions, while Pd-Pd ensembles catalyze C-C bond scission reactions. Density functional theory and microkinetic modeling show that the PdCu ensembles are bind CO fragments weakly, and hence destabilize C-C bond scission transition states. Similarly, the weaker CO binding onto low-index planes compared to stepped surfaces results in higher decarbonylation rates over smaller nanoparticles. Based on these results, we generalize our conclusions for a range of reactions, under oxidizing and reducing conditions, and present the implications of the segregation of alloy components to the catalyst surface on reactivity and selectivity.

References:

1. *Nature* 491 (2013) 235â€"239
2. *Am. Chem. Soc.* 138 (2016) 6805-6812
3. *Catal. Sci. Technol.* 8 (2018) 314-327

530475 Lubricant Base Oils Production from Biomass

Abstract:

The gradual depletion of world fossil fuel resources and global environmental concerns have prompted the use of biomass feedstock for renewable products. Lubricant is one such product that is primarily obtained from petroleum and it is introduced to reduce friction between surfaces in mutual contact. Base oils are the main components of lubricant and the majority of them are petroleum-derived mineral oils and synthetic oils (e.g. poly-Î±-olefins). The oils are formulated with different additives into several grades to suit automotive and industrial applications. Production of lubricant base oils from biomass will be an attractive alternative to petroleum derived-based oils to improve sustainability. So far, a few efforts have been devoted to the synthesis of lubricant-based oil alicyclic hydrocarbons, from biomass furans.[1-2] In this work, we report a novel route for selective production of base oil hydrocarbons from biomass platform chemicals and natural oils. The overall yield of hydrocarbons could reach up to 80%. Importantly, the hydrocarbon structures can be tailored using different biomass derived compounds to tune properties for various applications.

1. Balakrishnan, E. R. Sacia, S. Sreekumar, G. Gunbas, A. A. Gokhale, C. D. Scown, F. D. Toste and A. T. Bell, Proc. Natl. Acad. Sci. USA 2015, 112, 7645-7649.
2. Balakrishnan, G. E. Arab, O. B. Kunbargi, A. A. Gokhale, A. M. Grippo, F. D. Toste and A. T. Bell, Green Chem. 2016, 18, 3577-3581.

530529 Selectivity Control during the One-Pot Conversion of Aliphatic Carboxylic Acids to Linear Olefins through Tandem Hydrogenation/Dehydration

Abstract:

Olefins are important building blocks for the synthesis of higher value chemicals in the petrochemical industry and they are typically produced from fossil fuels. As an alternative to fossil fuel-based chemicals, biomass-derived carboxylic acids could serve as a viable replacement feed for olefin production. However, the production of olefins from carboxylic acids in one step remains a challenge due to the stability of carboxylic acids and the tendency of olefins to over-hydrogenate during hydrogenation, leading to alkanes.

In the present study, we have studied the one-pot catalytic conversion of linear aliphatic carboxylic acids to linear olefins via tandem hydrogenation/dehydration reactions. Hexanoic acid was converted to a mixture of hexenes over Cu nanoparticles supported on commercial silica-alumina (Siral 70) in a continuous flow reactor. At 483 K and 5 bar H_2 gauge pressure, we obtained 98.5% conversion of hexanoic acid, with a molar product distribution of 88.9% hexenes and 11.1% hexane. Remarkably, we observed a brusque selectivity switch to 99.9% hexane at 100% conversion.

A product intermediates study shows that when switching from a feed of pure hexene to a mixture of hexene and hexanoic acid, overhydrogenation of hexene can be suppressed. In addition, hexene temperature-programmed desorption (TPD) and FT-IR spectroscopy reveals that adsorption of hexanoic acid prevents hexene from accessing the catalyst surface, thus avoiding overhydrogenation. Reactions using butyric acid as feed also exhibit similar phenomena. We have also performed reactions utilizing butyric acid derived from biomass fermentation, albeit with significant catalyst deactivation.

We propose that the presence of a small amount of carboxylic acid on the surface of the catalyst prevents overhydrogenation of the olefins. In light of these results, current efforts are targeted at applying this concept to explain similar phenomena in other reactions, particularly in biomass conversion. The presence of numerous carboxyl groups in native/pre-treated biomass might explain the difference in catalyst performance/selectivity when processing model feed versus real biomass. Furthermore, the addition of carboxylic acids to the reaction medium during biomass upgrading could afford more control over the selectivity of the reaction.

526827 Mechanism and Kinetics of Isobutene Production over Zirconia-Supported Zinc Oxides

Abstract:

Growing interest in utilizing biomass-derived platform chemicals to produce fuels and specialty chemicals has motivated the investigation of the reaction of ethanol or acetone to form isobutene, a valuable specialty chemical used in the production of fuel additives, polymers, and other high-value products. Reports in the literature have identified zinc-zirconium mixed oxides as effective and selective catalysts for the production of isobutene from ethanol at 723 K. In this investigation, zinc-zirconium mixed oxides prepared via incipient wetness impregnation were selective and highly active catalysts for the acetone to isobutene reaction, reaching near-theoretical isobutene selectivities of over 86%. The mechanism and kinetics of the acetone and ethanol to isobutene reactions were studied in order to elucidate the reaction pathway, the roles of active acidic and basic sites, and the role of water in promoting stability and selectivity. Zinc-zirconium mixed oxide catalysts with varying zinc loadings were synthesized and characterized with XRD, BET, CO2-TPD, and DRIFTS FTIR with pyridine adsorption. Kinetic studies suggested that the ethanol to isobutene reaction proceeds via dehydrogenation to acetaldehyde, followed by condensation to form acetic acid, aldol condensation to produce acetone, dimerization of acetone to form diacetone alcohol, dehydration to form mesityl oxide, then finally hydrolysis of mesityl oxide to produce isobutene and acetic acid, which undergoes further aldol condensation to produce acetone. Understanding the reaction pathway, mechanism, and roles of the acidic and basic sites can help to define the catalyst properties required to selectively produce isobutene from a variety of sustainably-derived platform molecules.

533068 Catalytic Upgrading of Sugar-Derived Polyols to Glycols in Absence of Externally Added Hydrogen

Abstract:

Due to the increasing demand for fossil fuels and their diminishing supply worldwide, it is of considerable importance to utilize renewable resources for sustainable development of human society. Replacing fossil-based feedstock with biomass-derived ones to produce renewable fuels and chemicals is one of the promising ways to address challenges facing the energy and chemical industries. Sugar polyols, such as sorbitol, xylitol and erythritol are attractive platform chemicals because they can be readily derived from cellulosic biomass. Hydrogenolysis of sugar polyols can be converted to a variety of industrially important chemical intermediates including ethylene glycol (EG) and 1,2-propanediol (1,2-PDO), which are widely used as antifreezes, solvents and pharmaceuticals.

Up to date, it is known most existing hydrogenolysis processes are carried out under high temperature (T > 220 °C) and high H_2 pressure conditions (P > 4 MPa), which cause several technological issues such as poor product selectivity due to significant side reactions, fast catalyst deactivation as well as high capital cost and energy consumption.

In this work, we proposed a novel one pot H_2 generation and hydrogenolysis process in absence of external H_2 under mild conditions (T: 160 °C-200 °C, P: 1.0 Mpa nitrogen atmosphere).We have designed a series of bimetallic Pd-M (M: Cu, Ni, Pt, *etc*) catalysts supported on active carbon and layered double hydroxides and confirmed their remarkable performances in catalytic conversion of xylitol, sorbitol and mannitol to 1,2-PDO, EG and lactic acid (LA). In particular, catalyst activity and selectivity will be studied for different catalysts. Structure-performance correlations for H_2 generation and hydrogenolysis will be established based on bulk and surface characterization using XRD, SEM, TEM and chemisorption. The bimetallic Pd-Ni/AC and Pd-Pt/AC catalysts showed superior hydrogenated activities as well as selectivities towards lactic acid and glycols, compared with other bimetallic catalysts under identical conditions. At nearly 60.0% sorbitol conversion, a selectivity of 36.5% to ethylene glycol, propanediol and other glycols was obtained on Pd-Ni/AC at 200 °C, 1.0Mpa nitrogen atmosphere in absence of external H_2. Here we also researched reaction mechanism or pathways by studying metal active sites, H_2 spillover effect and the role of solid-base via regulating the growth of Pd-M and the ratio of acidity (basicity). It was concluded that H_2-spillover from Pd to M and solid-base acidity (basicity) may strongly influence the reaction mechanism to some extent, thereby influence the activity and selectivity of hydrogenolysis of biomass-derived to chemicals. Further details on reaction pathways and kinetic modeling will also be discussed in this presentation.

533584 Selective C-C Bond Scission of Primary Alcohols Using Cerium Oxide-Supported Palladium Catalyst

Abstract:

Biomass attracts significant attention as a carbon-neutral resource with the potential to suppress the CO_2 emissions. Recently, the biomass-derived chemicals such as glycerol, furfural, and levulinic acid are identified as the best suited alternatives for petroleum-derived chemicals. From the stand point of green sustainable chemistry, development of highly selective catalytic transformation of these compounds to valuable chemicals has been desired. To date, much effort has been devoted to selective C–O bond cleavage of high-oxygen containing biogenic polyols, such as glycerol, to produce valuable chemicals by hydrogenolysis and deoxydehydration.[1,2] On the other hand, C–C bond cleavage has not yet been widely researched despite its great potential to extend the utility of biomass-derived chemicals. The reported C–C bond cleavage methods often suffer from low selectivity toward the desired chemicals and limited substrate scope. In our group, we have reported the selective C–C bond scission of platform biomass-derived chemicals catalyzed by Ru/CeO_2 in the presence of water under 3 MPa of H_2.[3] In order to open new routes for industrially important chemicals from a wide range of biomass derivatives, development of highly selective and greener C–C bond scission catalysts is desired.

In this work, we found that cerium oxide-supported palladium (Pd/CeO_2) efficiently promoted the selective one-carbon scission of primary alcohols under air without H_2. The cleavage of C–C bonds occurs at positions adjacent to the hydroxymethyl groups (Scheme 1). The solid catalyst was reusable without appreciable loss of the activity and selectivity.

References

[1] T. Mizugaki, Y. Nagatsu, K. Togo, Z. Maeno, T. Mitsudome, K. Jitsukawa, K. Kaneda, *Green Chem.* **2015**, *17*, 5136–5139.

[2] T. Mizugaki, K. Togo, Z. Maeno, T. Mitsudome, K. Jitsukawa, K. Kaneda, *ACS Sustainable Chem. Eng.* **2016**, *4*, 682–685.

[3] T. Mizugaki, K. Togo, Z. Maeno, T. Mitsudome, K. Jitsukawa, K. Kaneda, *Sci. Rep.* **2017**, *7*, 14007.

536641 Electricity PerformanceÂ of Microbial Fuel Cell with Mixed Inoculation of Yeast and *E.coli*

Abstract:

Yeast and *E.coli* could be ideal MFC bio-catalyst since they are robust, safe, easily handled. What more they can metabolize a wide range of substrates with high catalytic rate. In this study, a typical two-chamber MFC with the only substrate of glucose was constructed by inoculation with yeast and *E.coli*. The results show that : The electric power of the mixed inoculated MFC (351.45mW/ m^2) is obviously better than that of the single inoculated MFC. This Highest electric power is 47% higher than *E.coli* based MFC (185.71 mW/m^2) and 39% higher than that of yeast based MFC (215.22mW/m^2). The internal resistance of MFC is smaller when it inoculated with both yeast and *E.coli* rather than a single organism, which means these two organisms play a mutually reinforcing role in the process of electron transfer. Therefore, mixed inoculation of yeast and E.coli in MFC reduce resistance of electron transfer as well as whole resistance of MFC.

534173 Biological Detoxification of Lignocellulosic Biomass Hydrolysate Liquor for Enhanced Ethanol Production

Abstract:

Acid pretreatment of lignocellulosic feedstock during biofuel production leads to the formation of toxic by-products, such as furfural and 5-hydroxy-2-methylfurfural (HMF), which inhibit the downstream enzymatic hydrolysis and fermentation. An efficient and environment friendly process for removal of these inhibitors is essential. The present study reports the isolation of a bacterium, *Bordetella* sp. BTIITR, from soil that exhibited capability to selectively degrade furfural and HMF without significant consumption of sugars (1). This substrate priority of *Bordetella* sp. BTIITR for HMF and furfural makes it a potential organism for biodetoxification of lignocellulosic hydrolysate liquor. The isolated bacterium was able to degrade 100 % of furfural and 80 % of HMF from simulated hydrolysate liquor in 16 h of incubation period. For the actual sugarcane bagasse hydrolysate liquor, it was able to remove 100 % of furfural, 94 % of HMF and 82 % of acetic acid in 16 h of incubation period. Further, *Bordetella* sp. BTIITR was immobilized within chitosan beads and the beads were utilized to detoxify the hydrolysate. The immobilized cells were able to degrade 94 % HMF and 100 % furfural in 20 h at pH 8 and 35- 40 ^{0}C temperature. The detoxified hydrolysate liquors demonstrated improved fermentation efficiency for production of ethanol as compare to the undetoxified hydrolysate. Isolation of bacteria, characterization, detoxification in free and immobilized forms, and effect on ethanol production will be discussed. This is the first study to report the capability of *Bordetella* to metabolize the furan derivatives.

Reference: 1. B. Singh, A. Verma, Pooja, P. K. Mandal, **S. Datta**[*], A Biotechnological Approach for Degradation of Inhibitory Compounds Present in Lignocellulosic Biomass Hydrolysate Liquor Using Bordetella sp. BTIITR, *Chem. Eng. J.*, **328** (2017) 519-526

534555 From Starch-Enriched Algal Biomass to Biobutanol Production – a Model-Based Optimisation Study

Abstract:

Biobutanol is a sugar-based biofuel produced by the ABE fermentation, an old biochemical process whereby sugars are metabolised into acetone (A), butanol (B), and ethanol (E). The commercial implementation of the ABE fermentation, however, is restricted due to: i) low butanol fermentation yields and ii) the use of common feedstocks which raise *food vs fuel* concerns (food-based biomass) or compete for arable land (lignocellulosic biomass) (Kumar and Gayen, 2011). Microalgal biomass has the potential to become a sustainable large-scale feedstock suitable for the ABE fermentation due to its ability to accumulate starch (Markou et al., 2012). Nutrient-stressed algal cultivation has been shown to induce starch accumulation, but such a strategy must be optimised to sustain algal growth while simultaneously enhancing starch formation.

In order to establish the best microalgae-to-biobutanol route, we have previously employed a kinetic model capable of accurately predicting algal growth dynamics subject to mixotrophic growing conditions co-limited by nitrogen and phosphorus (Figueroa-Torres et al., 2017). The proposed model was successfully validated against datasets obtained from lab-scale cultures of *Chlamydomonas reinhartii CCAP 11/32c*. In this work, "starch-enriched" algal biomass, as obtained from a validated model-based cultivation strategy maximising starch formation, has been evaluated as a feedstock for biobutanol production. Thus, a kinetic model has been developed in combination with a range of lab-scale batch fermentation experiments with the wild-type strain *Clostridium acetobutylicum* DSM 792 (at 37 °C under anaerobic conditions) to: i) identify optimal pre-treatment steps required to use "starch-enriched" microalgal biomass as a fermentable feedstock, ii) establish optimal operating conditions for biobutanol production, and ii), and establish the kinetics of biobutanol production from "starch-enriched" algal biomass.

References

Figueroa-Torres, G.M., Pittman, J.K., Theodoropoulos, C., 2017. Optimisation of microalgal starch formation for the biochemical production of biobutanol. Comput. Aided Chem. Eng. 40, 2899–2904.

Kumar, M., Gayen, K., 2011. Developments in biobutanol production: New insights. Appl. Energy 88, 1999–2012.

Markou, G., Angelidaki, I., Georgakakis, D., 2012. Microalgal carbohydrates: an overview of the factors influencing carbohydrates production, and of main bioconversion technologies for production of biofuels. Appl. Microbiol. Biotechnol. 96, 631–645.

532270 Analysis of the Topochemistry of Lignocellulosic Biomass and Modeling of the Reaction Dissolution Phenomena

Abstract:

Successful incorporation of plant-based materials into biorefinery processes is heavily dependent pretreatments which allow economically-sensible access to target compounds. Confocal Raman Microscopy (CRM) offers the unique analytical opportunity to - without any chemical specimen alteration - qualitatively and semi-quantitatively observe the *in muro* spatial distribution of structural components. The effect of hot water pretreatment on the distribution of lignin and cellulose inside the cell walls of *Acer saccharum* was studied. Wood chips were transversely microtomed into thin sections of 10 Âµm. These sections were pretreated in hot water at 160 C for varying lengths of time and then analyzed by CRM. Alternatively, wood chips were pretreated first and then subject to CRM analysis. Raman spectra were obtained by excitation with a polarized laser (Î»=532nm) on a LabRam Xplora Confocal Raman Microscope. Spectra were analyzed by the equipmentâ€™s built-in LabSpec software component. Primary component and cluster analyses of the data were also performed to sort data corresponding to morphologically distinct cell wall layers.

The results showed for the first time, that lignin dissolved progressively from the secondary walls and the compound middle lamella regions. Samples where the lignin dissolution was limited by transport within the cell walls were markedly heterogeneous and showed large variations in distributions within the morphologically distinct CML and secondary layers. On the other hand, when thin section samples were subjected to pretreatments, the dissolution of lignin was more uniform, and clear distinctions between the morphological layers were observed. CRM distributions of these cell wall components can provide good data for understanding and modeling the physio-chemical processes occurring in the cell walls under different pretreatment conditions.

536852 Biomass Torrefaction in a Pulsed Fluidized Bed

Abstract:

Torrefaction at 200-300°C in an inert or oxygen-deficient environment is a promising pre-treatment for improving the properties of biomass for energy utilization. Torrefied biomass has a higher heating value, increased hydrophobicity and better grindability.

Common reactor technologies adopted for biomass torrefaction are fixed bed, moving bed, rotary drum and auger type reactor. However, none of these technologies really stands out, because of associated problems of solid mixing, controllability, inefficient heating etc. Nevertheless, aforementioned technologies treat biomass at a small heating rate, which requires a large volume to achieve a commercial capacity. Fluidized bed on the other hand has enhanced heat and mass transfer, excellent uniform heating and mixing. A few applications of fluidized bed for biomass torrefaction have been studied. However, unconventional nature of biomass particles such as wide particle size distribution, low density and irregular shape is problematic and even causes defluidization. To improve the fluidization quality, inert materials are added to the bed. Extra efforts on separation make fluidization with bed materials unfavourable.

Another technique to assist biomass fluidization is gas pulsation. Biomass drying in a bench scale pulsed fluidized bed reactor has been performed and studied in our group. Based on the knowledge gained from the previous work, a novel pulsed fluidized bed (PFB) reactor for biomass torrefaction, with continuous feed and discharge, has been designed and commissioned in the presented study. The PFB unit has successfully demonstrated for sawdust torrefaction. Furthermore, effects of gas pulsation frequency, torrefaction temperature, biomass feed rate and oxygen concentration in carrier gas on torrefied product properties are being investigated. The latest results on suitable operating conditions for biomass torrefaction in PFB unit will be presented. This new torrefaction technology will promote the utilization of local biomass resources for displacing non-renewable fossil fuels.

534276 Co-Hydrothermal Carbonization of Coal Refuse and Cow Manure Blend

Abstract:

Coal refuse, also known as coal waste and gob, are by products resulted from coal mining. Coal processing plants rejects 30-60% materials from the mined coal during the cleaning procedure and it contains 40-70% energy compared to coal. However, toxic elements (sulfur, heavy metals and ash) present in coal refuse possess serious environmental threat and that is why it is not possible to combust coal refuse without prior treatment. That is why, it is necessary to perform a pre-treatment process to reduce the amount of toxic elements before using coal refuse.

Co-hydrothermal treatment (co-HTC) of coal refuse along with a biomass (e.g. cow manure, pine wood or Miscanthus) can be an efficient option to utilize this huge amount of available fuel resource. Co-HTC is a thermochemical process, where feedstock is treated at high temperature (180-280 °C) and corresponding vapor pressure (5-60 bars). During the process, biomass will provide additional carbon and organic substances to coal refuse and produce final product of similar fuel properties of coal. Previous studies have shown significant decrease of toxic materials when hydrothermal treatment was performed on coal and biomass mixture.

This study will perform co-HTC of coal refuse and cow manure at three different temperatures (180, 230 and 280°C) for 10, 30 and 50 min of residence time for each temperature with a 50-50 (dry wt. basis) coal refuse-biomass ratio. Fuel quality of the product of co-HTC process will be determined by evaluating mass yield, energy content, ultimate analysis and proximate analysis. IC and SEM-EDS analysis of solid product will be performed to analyze the level of toxic compound concentration.

535228 Advanced Characterization of Poplar Variants for Understanding Plant Cell Wall Recalcitrance

Abstract:

Certain poplar variants have been identified to carry a rare natural mutation associated with EPSP (5-enolpyruvylshikimate-3-phosphate) synthase biosynthesis gene that lowers their lignin content. Our recent work has shown that the rare variant (BESC-35) and low lignin comparator (GW-11012), had reduced klason lignin content of 18%. On the other hand, a high lignin comparator (BESC-316), had a klason lignin content of 23%. Rare variant and low lignin comparator were some of the highest-ranking plants and high lignin comparator was one of the low-ranking plants, with respect to sugar yield from pretreatment and enzymatic hydrolysis combined. In this study, differences in cell wall properties of these three high or low sugar yielding poplar plants were investigated using SANS, AFM, TEM, NMR, GPC, and compositional analysis. SANS suggested that the rare variant had looser packing of cellulose microfibrils. AFM using PeakForce quantitative nanomechanical property mapping (QNM) found that the high lignin comparator had significantly higher tip-surface adhesion than rare variant or low lignin comparator possibly due to its higher lignin content. Moreover, lower syringyl/guaiacyl (S/G) ratio of high lignin comparator (2.58) agreed with the assertion that lower S/G ratio in poplar can negatively affect sugar yield. However, despite the rare variant and low lignin comparator being low in lignin content and high sugar yielding plants, their S/G ratios of 4.27 and 3.13 were widely different. AFM and TEM suggested high heterogeneity in cell walls of high lignin comparator. Overall, these findings provide new insights into differences in poplar plants associated with recalcitrance.

534301 Simulation of CO_2 Gasification of Manure-Derived Hydrochar Using Aspen Plus

Abstract:

Approximately 160 million tonnes of cow manure are produced every year in the US. However, high moisture content, eutrophication, greenhouse gas emissions, and excreted pathogens are some major concerns associated with livestock manure, which limit the manure application as fuel and fertilizer. Direct combustion of manure is not an efficient option due to the high moisture content (>80 %) and can produce large amounts of SO_x and NO_x. Therefore, an efficient energy conversion process is necessary to utilize this large amount of annually accumulated livestock manure for bioenergy generation.

Gasification of hydrothermally treated cow manure can be an efficient option to utilize this alternate source of energy. Gasification process will convert the hydrothermally treated feedstock into a mixture of gaseous products in presence of limited amount of oxidizing agents like oxygen (O_2), steam, air or carbon dioxide (CO_2) or a mixture of these agents. The gaseous product will consist of mainly carbon monoxide (CO) and hydrogen (H_2), known as synthetic gas or syngas which can be converted to different fuel forms as necessary. This process also reduces to amount of CO_2 produced which makes it environmental friendly.

The goal of this study was to simulate the gasification process in ASPEN plus V9 simulation software using previously determined reaction kinetics and CO_2 as the oxidizing agent to evaluate the effect of process parameters (temperature, equivalence ratio) on syngas production. Results showed that CO and H_2 concentration increased with increasing temperature and equivalence ratio. However, after a certain range the CO concentration started decreasing. These results were later validated using data from a practical gasification experiment performed in a downdraft gasifier.

522625 Biomass Residue Characterization for Their Potential Application As Biofuels

Abstract:

Abstract

Physico– and thermo– chemical characteristics of four types of walnut shells (PSW, TSW, MSW and HSW) have been investigation in the present work. From the proximate analysis (wt% on dry ash free basis), the highest moisture, volatile matter, ash and fixed carbon contents is observed in PSW (9.26), HSW (79.8), MSW (5.84) and TSW (26.1), respectively. It is found that TSW have lowest moisture (8.08) and volatile matter (73.9) and HSW has lowest ash (2.0) and fixed carbon (20.2) contents. From the ultimate analysis it is found that the H/C and O/C ratios are almost 1.6 and 0.8, respectively, for all types of walnut shells. Heating values are observed in the range of 13.8–18.4 MJ/kg, which is comparable to the wood waste and lignite coal. The cellulose, hemicellulose, lignin and extractives in walnut shell are found to vary from 32.3-34.5, 21-27, 39-43 and 1.4-1.7%, respectively. From the thermogravimetric analysis (TGA), it is observed that the complete moisture removal is below 152 ^{o}C. The highest and lowest moisture losses of 9 and 5% have been observed at a heating rate of 5 °C/min for MSW and TSW, respectively. The degradation of lignocellulosic biomass is occurred between 330 to 510 °C. For all walnut shell samples the most prominent FTIR band peak have been found at wavenumbers of 3400, 2931, 1420 and 1050 cm^{-1}, which is due to the stretching vibrations of –OH, CH–, aromatic C=C, and aliphatic ether and alcohol groups, respectively in cellulose, hemicellulose and lignin. Scanning electron microscopy (SEM) analysis indicated the rough texture and heterogeneous structures of biomass. Further, the x-ray diffraction (XRD) analysis showed the crystalline structure, which is due to the presence of cellulose. From the observations made in the present work, it can be concluded that the walnut shell is a potential candidate for energy generation through thermo-chemical conversion.

Keywords: Walnut shell, Proximate and Ultimate Analysis, Higher heating values, Compositional analysis

524736 Gas Phase Catalytic Oxidation of Lignin to Produce Phenolic Compounds over Vanadia Catalysts

Abstract:

The pretreatment of lignocellulosic biomass to create fermentable sugar streams *en route* to fuel/chemical production may also produce streams of insoluble lignin. These lignin stream can be burned to produce process heat and power. An alternative use of the lignin streams could be the production of valuable products such as phenolic compounds. This work explores the pyrolysis of lignin residue to produce gas phase products which can be passed of vanadia-based catalysts to oxidatively remove methoxy side chains to produce phenolic compounds, which could be used in the production of polymers and resins. Because lignin contains a complex mixture of linkages with methoxy groups, as well as both C-C and C-O bonds linking phenyl groups, there are number of bonds that must be broken in order to funnel lignin into the desired slate of phenolic compounds without methoxy groups (simple phenols). The subject of this presentation uses pyrolysis of lignin to create a mixture of gas phase species that can be passed over a catalyst to oxidatively cleave the methoxy groups to produce simple phenols. Because the goal is to partially oxidize the lignin pyrolysis vapors to produce phenols, as opposed to over-oxidation which would result in CO and CO_2 production, the oxidation activity of the catalysts must be controlled. Catalysts with varying vanadium-oxygen bond strength, which affects the oxidation activity, were designed by altering the catalytic support and synthesized using incipient wetness impregnation. Reaction experiments were conducted by passing lignin pyrolysis vapors over the catalysts at 500 °C and were shown to produce simple phenols from methoxy phenols. The effect of catalyst composition will be discussed along with relevant characterization techniques, coupled with computational modeling, to help understand the role of metal-oxygen bond strength in the catalytic activity.

526393 Mechanistic Study of the Hydrogenolysis of Diaryl Ethers Catalyzed By Heterogeneous Metal Catalysts

Abstract:

Manipulating reaction pathways of metal-catalyzed reactions is key to efficiently convert polar aromatic compounds. Diaryl ethers undergo C-O hydrogenolysis and hydrogenation with rates that depend on the specificity of the metal. We are studying the corresponding catalytic chemistry of Ni, Pd, Ru and Pt with reaction kinetics and experiments with isotopically labeled reactants. We have found a common reaction mechanisms for all studied metals but different isotopic effects, which points to the interaction of hydrogen with the metal as the factor determining selectivity. The details of the mechanisms, which are being tested by theoretical calculations, will be reported and selectivity rationalized as a function of the binding energy of reactants with the metal.

527899 Palladium-Iron Bimetallic Catalyst: High Activity and Stability for Aqueous Phase Hydrogenations

Abstract:

A novel palladium-iron bimetallic catalyst was synthesized that provided high hydrogenation activity and stability relative to palladium only under harsh hydrothermal conditions. To generate the catalyst, palladium was dispersed on carbon coated silica, which was doped with iron. The catalyst was applied to the hydrogenation of carbonyl groups in acetone, 2-pentanone, and propionaldehyde. The Pd-Fe catalyst gave more than a three-fold higher hydrogenation activity than the nanoparticle Pd only counterpart, while Fe nanoparticles were shown to be inert. The synthesis procedure was optimized for the bimetallic material, which demonstrated that the introduction of Fe contributed to a higher turnover frequency rather than simply increases the number of active sites. Interestingly, the same synthesis procedure was used to synthesize a Pd-Mo bimetallic catalyst that also increased activity.

536491 Hydrodeoxygenation of Guaiacol over Ni and Mo Nanoparticles Supported on SBA-15 and γ-Al_2O_3.

Abstract:

1. **Introduction**

The limitation of fossil energy resources and the increasing need for chemicals and liquid fuels require strong efforts in utilizing renewable energy sources. In this way, a technology such as biomass to liquid fuels (BTL) is being focus. The fast pyrolysis is a promising tool to transform various biomass feedstock's, lignocellulosic biomass, into an energy-dense and oily mixture. However, it possesses considerable amounts of oxygenates molecules, phenolic compounds and water in the material obtained by this method make it impossible to use the final product as motor fuel. For the use in chemical industry or fuel production an upgrading of the pyrolysis oil is necessary, which can be achieved via hydrodeoxygenation (HDO). This process consists in the treatment of bio-oil at temperatures between 200 and 450 °C with hydrogen at high pressure (10–300 bar) in the presence of a catalyst, so that the oxygen contained in organic molecules can be extracted with water formation[1]. Research efforts to study the chemistry of hydrodeoygenation of biomass-derived oil are gaining considerable importace over the last 20 years[1].The most frequently examined HDO catalysts commercially used for petroleum hydrotreatment products are NiMo/Al_2O_3 and CoMo/Al_2O_3.In the present study catalysts based on supported Ni and Mo are evaluated as catalysts for HDO reaction. The effect of metal phase and support properties, SBA-15 and γ-Al_2O_3 (reference support), as well as their interactions has been investigated. For the catalytic reactions, guaiacol was chosen as a model compound for lignin depolymerization fragments formed during bio-oil pyrolysis from lignocellulosic biomass.

2. **Methods**

The pure mesoporous silica SBA-15 was synthesized by hydrothermal method. The γ-Al_2O_3 was purchase from Degussa. The bimetallic nickel-molybdenum catalysts were prepared by wet impregnation. First, appropriate quantities of $(NH_4)_6Mo_7O_{24}.4H_2O$ and $Ni(NO_3)_2.6H_2O$ were dissolved in deionized water; the support was added and stirred for 2 h. Then, the suspension was freeze dried and the catalyst was calcined at 500 C for 2 h, under air flow. The metal contents were 15% of MoO_3 and 5% de NiO (weight percentage) with relation to support. The materials were characterized by nitrogen physisorption isotherms, transmission electron microscopy (TEM), X-ray powder diffraction (XRD), energy-dispersive spectroscopy (EDS) and with Raman and infrared spectroscopy. The total acidity of the catalysts was measured by Temperature programmed desorption of NH_3 (TPD-NH_3). The catalytic tests were carried out in a fixed bed flow reactor system under 15 kgf/cm^2 pressure. The catalysis, mass around 200 mg, was pre-treated in situ with 50 cm^3/min flow of pure H_2 at 500° C for 4 h and then cooled to the reaction temperature (200, 250 or 300° C). The feed mixture consisted of 2% w/w guaiacol (MERCK) in heptane P.A. (Aldrich) with 14.8

cm^3/h flow and 100 cm^3/min H_2 flow. The feed and the products were analyzed by gas chromatography Shimadzu CG17A with FID detector, capillary column CPsil 5CB and automatic injection.

3. Results and discussion

The fresh catalysts X-Ray diffractogram show that the Ni and Mo were successfully incorporated under the SBA-15 and Al_2O_3 supports as NiO and MoO_3 nanoparticles. After nickel and molybdenum incorporation the surface area of the SBA-15 decrease from 630 to 332 $m^2.g^{-1}$. The same behavior was observed for γ-Al_2O_3 support, the surface area chance from 279 to 233 $m^2.g^{-1}$. The temperature reduction of Ni and Mo oxide were determined by TPR-H_2, 500 °C under H_2 atmosphere for 4 hours. The catalysts total acidity for NiMo/SBA and NiMo/ALU were 470 and 548 μmol.g^{-1}, respectively. The results of catalytic activity for both materials are summarized in Table 1. For NiMo/SBA and NiMo/ALU the reaction temperature have a positive effect in guaiacol conversion, the higher results were observed at 300 °C 84.7% and 72.6% for NiMo/SBA and NiMo/ALU, respectively. However, the higher HDO values occurred for reaction temperature at 200°C. Thus, the selectivity for deoxygenated compounds were more pronounced in low temperatures for both catalysts. The principal products formed during the guaiacol HDO reaction were benzene, cyclohexane, cyclohexene, phenol and o-cresol. Compared to NiMo/ALU the NiMo/SBA showed the best performance in HDO, HDA and higher selectivity for benzene in all reaction temperature studied.

Table 1: The catalytic results of NiMo/SBA and NiMo/ALU in the hydrodeoxygenation of guaiacol.

	NiMo/SBA			NiMo/ALU		
Temperature (°C)	**200**	**250**	**300**	**200**	**250**	**300**
Conversion (%)	3.7	74.5	84.7	5.1	45.0	72.6
HDO (%)	92.1	74.3	66.5	37.9	31.5	30.8
HDA (%)	84.2	48.7	35.3	23.1	12.3	2.7
Selectivity (%)						
Benzene	16.7	8.6	1.3	3.3	1.1	0.3
Cyclohexane	29.1	13.9	11.0	6.8	3.3	0.4
Cyclohexene	38.5	26.2	19.5	12.9	7.9	1.9
Phenol	10.5	17.4	34.2	16.3	24.8	41.3
O-Cresol	0.0	18.4	8.0	7.7	8.5	11.3

With the catalysts thermogravimetric analysis were possible quantified the coke formation after HDO reactions. For NiMo/ALU and NiMo/SBA the coke formation were 1,65 and 1.14 $mg.g_{cat}^{-1}.h^{-1}$.

4. Conclusions

Compared with NiMo/ALU the NiMo/SBA show the best performance in guaicol conversion, selectivity for deoxigenated compound and in HDO percentage. The principal compounds formed during the HDO reaction with NiMo/SBA were cyclohexene, cyclohexane and bezene. For NiMo/ALU the principal compounds were phenol and o-cresol. Thus, the mesoporous silicon oxide (SBA-15) was the best support for nickel e molybdenum nanoparticles resulting in the catalyst for guaicol hydrodeoxygenation.

References

- Arun, R.V. Sharma, A.K. Dalai, Renew. Energy Rev., 48 (2015) 240-255.

Keywords

Hydrodeoxygenation, catalysis, nickel, molybdenum.

530385 Hydroprocessing of Biomass-Derived Oxygenates on Metal-Exchanged Zeolites Using Light Alkanes As the Source of Hydrogen

Abstract:

The broader use of natural gas and biomass for energy applications is limited because of the short hydrocarbon chains and/or high oxygen content of the molecules that comprise these resources. Current technologies for natural gas and biomass upgrading require high temperatures and suffer from poor selectivities. Furthermore, biomass deoxygenation requires external sources of H_2 molecules that are most commonly derived by sacrificing a portion of the feedstock in steam reforming reactions. In the absence of molecular hydrogen, a portion of the carbon feedstock is converted to arenes during hydrogen transfer reactions to satisfy the stoichiometry for oxygen removal as water. This work aims to improve selectivity and solve the hydrogen deficiency issue by developing new catalysts that can facilitate hydro-deoxygenation reactions using light alkanes as the hydrogen source. These hydrogen transfer steps have the potential to selectively terminate growing hydrocarbon chains via desorption as less reactive alkanes, thereby producing targeted classes of compounds, as opposed to a wide range of products over a broad carbon number distribution. Furthermore, this strategy has the potential to upgrade a portion of the light alkane feed because the hydrogen transfer steps generate alkenes that can participate in subsequent C-C bond formation reactions.

Reactions of isobutane and n-butanal (either as single reactants or co-feeds) on H-BEA zeolite and H-BEA samples partially exchanged with Zn cations were studied to probe the effects of alkanes as co-reactants during acid-catalyzed reaction of oxygenates. The Zn-exchanged BEA zeolite was synthesized via ion-exchange using aqueous solutions of zinc nitrate followed by treatment in flowing, dry air first at 373 K for 20 h then at 773 K for 4 h. This method has been previously shown to effectively exchange protons for Zn^{2+} ions within the cages of zeolites. n-Butanal reactions (2 kPa) on H-BEA predominantly form 2-ethyl-2-hexenal via aldol-condensation (60% selectivity to the aldol-condensation dimer) while also forming xylenes and cyclic ketones via cyclization/deoxygenation of 2-ethyl-2-hexenal (25% selectivity). n-Butene formation was also observed (selectivity of 15%) via hydro-deoxygenation of n-butanal. Addition of isobutane as a co-feed on H-BEA leads to minor changes in the product selectivity, indicating that the unsaturated species formed from n-butanal conversion are not strong enough hydride acceptors to activate isobutane via hydride transfer on Bronsted acid sites. Product selectivities for n-butanal conversion on Zn-H-BEA were similar to those on H-BEA, suggesting a minor effect of the Zn sites on n-butanal reaction paths. Co-reaction of isobutane (40 kPa) and n-butanal (2 kPa) on Zn-H-BEA exhibited a completely different product distribution. The predominant products were C_4 hydrocarbons (34% selectivity with 50% selectivity to n-butane within C_4) and C_8 alkanes (dimethyl-hexane and methyl-heptane isomers; 31% selectivity). n-Butane is the expected isomer formed from deoxygenation-hydrogenation of n-butanal, and the C_8 isomers that were produced were those that would be expected from deoxygenation of aldol-condensation (methyl-heptanes) and Prins-condensation (dimethyl-hexanes) products. Reaction of isobutane/n-butanal feeds on the Zn-H-BEA catalyst also exhibited lower selectivity to the aldol-condensation dimer (16% versus 70%) and to cyclic species (16% versus 25%) compared to isobutane/butanal reactions on H-BEA. Isopentane formation (5% selectivity) and C_7 alkanes (methyl-hexanes and n-heptanes; 9% selectivity) were also observed on Zn-H-BEA during conversion of isobutane/butanal feeds. These C_7 isomers are what would be expected from deoxygenation of C_7 ketones derived from Tishchenko esterification-ketonization routes. Reaction of isobutane (40 kPa) was also carried out on H-BEA and Zn-H-BEA to identify the extent of isobutane dehydro-oligomerization reaction paths. Isobutane conversion was only 0.1% on H-BEA and only 0.2% on Zn-H-BEA, indicating that dehydro-oligomerization of isobutane only occurs to a negligible extent in parallel with n-butanal conversion paths. This negligible conversion of isobutane is consistent with the absence of trimethyl-pentane isomers from the product distributions (indicating undetectable isobutane dehydro-oligomerization).

530346 Flowthrough Reductive Catalytic Fractionation of Biomass

Abstract:

As an abundant and renewable source of carbon, lignocellulosic biomass is an excellent feedstock for sustainable fuel and chemical production. Biomass primarily consists of carbohydrates (cellulose and hemicellulose) and lignin. Carbohydrates have been extensively studied and shown to be versatile feedstocks for the production of fuels and chemicals. Lignin, the largest source of natural aromatics, becomes highly recalcitrant during carbohydrate processing, hindering chemical production from lignin. Reductive catalytic fractionation (RCF) has emerged as a leading biomass fractionation and lignin valorization strategy that simultaneously extracts, depolymerizes and stabilizes lignin without disturbing the carbohydrate fractions. Here, flowthrough reactors were used to investigate RCF of poplar. Most RCF studies to date have been conducted in batch, but a flow-based process enables the acquisition of intrinsic kinetic and mechanistic data essential to accelerate the design, optimization, and scale-up of RCF processes. Time-resolved product distributions and yields obtained from experiments with different catalyst loadings were used to identify and deconvolute events during solvolysis and reductive stabilization. Multi-bed RCF experiments provided unique insights into catalyst deactivation, and initial rate studies enabled the determination of kinetic parameters intrinsic to the solvolytic step of RCF. Overall, this work demonstrates the concept of flowthrough RCF, which will be vital for realistic scale-up of this promising approach.

530547 Hydrodeoxygenation of Sorbitol to Monofunctional Fuel Precursors over Co/TiO_2

Abstract:

Over the next two decades, demands for heavier fuels such as jet and diesel are projected to increase while demands for lighter fuels such as gasoline are projected to decrease.[1] Production of these heavy C_8-C_{22} hydrocarbons remains a challenge when using biomass feedstocks in large part because the monomeric units which comprise cellulose and hemicellulose possess only five or six carbon atoms. One potential route for producing these fuels involves converting carbohydrates into C_{5+} monofunctional oxygenates (MFs) by aqueous-phase hydrodeoxygenation (APHDO) followed by performing C-C coupling reactions.[2] APHDO consists of feeding a highly-oxygenated species in water at moderate temperatures (~500 K) along with hydrogen at high pressures (~6 MPa) over a bifunctional catalyst usually utilizing a noble metal on an acidic support.[3] In this chemistry it is desired to control the cleavage of C-O bonds and minimize the cleavage of C-C bonds by retro-aldol condensation, decarbonylation and decarboxylation reactions.[4] However, over-dehydration to lower-demand alkanes is undesired. In addition, the noble metals commonly used are costly, but utilization of low-cost base metals is often limited in APHDO due to their propensity to leach into the aqueous phase. However, recent developments have shown that Co/TiO_2 catalysts can be stabilized for aqueous-phase hydrogenations utilizing high-temperature pretreatments which take advantage of the strong metal-support interaction (SMSI).[5] The purpose of this study is to investigate the potential for MF production in sorbitol APHDO using a SMSI-stabilized Co/TiO_2 catalyst.[6]

Sorbitol APHDO of Co/TiO_2 produced a 56% yield (on a carbon basis) to MFs at a moderate WHSV of 0.70 h^{-1}. This is the highest yield reported from sorbitol in the literature. We estimate the Co catalyst is 50 to 1,000 times cheaper than previously reported precious metal catalysts. Increasing space time leads to excessive deoxygenation and therefore selectivity toward alkanes and CO_x while decreasing space time leads to higher oxygenates, including diols, oxygenated heterocycles, unidentified O_{3+} species. The MFs produced are mostly primary alcohols (37% at 0.70 h^{-1}), secondary alcohols (32%), and heterocycles (23%). Ketones (7%) and aldehydes (<1%) are produced in much lower yields. 67% of the MFs are C_{5+} and would therefore be suitable for coupling. FTICR-MS shows that sorbitol/anhydrosorbitol species can oligomerize at lower contact times to form C_9-C_{18} species which appear to break down to some degree at longer contact times to eventually form MFs. Overall, the product distribution vs. WHSV suggests that about one third of the fed carbon reacts through the retro-aldol route and overall 20% of the carbon is lost to C_1 gases such as CO_2 and methane. The need to suppress retro-aldol condensation and decarbonylation in sorbitol APHDO is a clear limitation for the production of C_{5+} MFs. In the present study, Co/TiO_2 catalyst shows irreversible deactivation with TOS due to sintering and leaching of the Co. Overall, this work shows that Co/TiO_2 can be used to produce monofunctional fuel precursors from sorbitol at high yields, but suffers from irreversible leaching and sintering due to chelation from highly-oxygenated species.

References

1. ExxonMobil, 2017 Outlook for energy: A view to 2040. ExxonMobil: Irving, 2017.
2. Kunkes, E. L.; Simonetti, D. A.; West, R. M.; Serrano-Ruiz, J. C.; Gärtner, C. A.; Dumesic, J. A., Catalytic conversion of biomass to monofunctional hydrocarbons and targeted liquid-fuel classes. *Science* **2008,** *322* (5900), 417-421.
3. Vilcocq, L.; Cabiac, A.; Especel, C.; Guillon, E.; Duprez, D., Transformation of sorbitol to biofuels by heterogeneous catalysis: Chemical and industrial considerations. *Oil & Gas Science and Technology–Revue d'IFP Energies nouvelles* **2013,** *68* (5), 841-860.
4. Li, N.; Huber, G. W., Aqueous-phase hydrodeoxygenation of sorbitol with Pt/SiO_2–Al_2O_3: Identification of reaction intermediates. *Journal of Catalysis* **2010,** *270* (1), 48-59.
5. Lee, J.; Burt, S. P.; Carrero, C. A.; Alba-Rubio, A. C.; Ro, I.; O'Neill, B. J.; Kim, H. J.; Jackson, D. H.; Kuech, T. F.; Hermans, I.; Dumesic, J. A.; Huber, G. W., Stabilizing cobalt catalysts for aqueous-phase reactions by strong metal-support interaction. *Journal of Catalysis* **2015,** *330*, 19-27.
6. Eagan, N. M.; Chada, J. P.; Wittrig, A. M.; Buchanan, J. S.; Dumesic, J. A.; Huber, G. W., Hydrodeoxygenation of Sorbitol to Monofunctional Fuel Precursors over Co/TiO2. *Joule* **2017,** *1* (1), 178-199.

533566 Catalytic Hydrogenation of Furfural over Rumop: Probing Bimetallic andÂ Compositional Effects on Catalyst Performance

Abstract:

Catalytic Hydrogenation of Furfural over RuMoP: Probing Bimetallic and Compositional Effects on Catalyst Performance

Yolanda Bonita, Timothy P. Oâ€™Connell, and Jason C. Hicks*

Department of Chemical and Biomolecular Engineering, University of Notre Dame, 182 Fitzpatrick Hall, Notre Dame, IN 46556

The synthesis of furfuryl alcohol from furfural uses an environmentally toxic copper chromite catalyst in the process. As possible alternatives, Cu-based catalysts such as supported Cu, bimetallic Cu-Ni, Cu-Co, Cu-Fe, and Cu-Ru have been studied to replace copper chromite. However, none of these catalysts surpass the high selectivity exhibited by copper chromite for furfural hydrogenation to furfuryl alcohol.

We have recently discovered a variety of novel transition metal phosphides that are highly selective to furfuryl alcohol and active at near ambient temperature. Contrary to the literature that suggested group VIII metals are not selective to furfuryl alcohol due to the binding configuration, we have synthesized group VIII metal (i.e. Ru) phosphides that exhibit high selectivity to furfuryl alcohol, similar to that of copper chromite and higher than other Cu-based catalysts. Moreover, when comparing pseudo-first order rate constants, we see an increase of > 5 times in the site-normalized, initial rate and 99% selectivity for bimetallic RuMoP. This phenomena was also observed in our previous studies (i.e. hydrodeoxygenation reaction)[1-4], which showcases bimetallic character and potential control of the catalytic properties based on catalyst composition. Indeed, we have observed direct influence on the activity and selectivity by varying the composition of $Ru_xMo_{2-x}P$, where the measured pseudo-first order rate constant was enhanced for Ru-rich materials ($x = 1.2$). In this presentation, we will present a thorough examination of bimetallic RuMoP as a furfural hydrogenation catalyst through detailed characterization and kinetic experiments.

References

1. Rensel, D. J., Rouvimov S., Gin M. E., Hicks J. C. Highly selective bimetallic FeMoP catalyst for C-O bond cleavage of aryl ethers. *J. Catal.* Sep 2013;305:256-263.
2. Rensel, D. J., Kim J., Bonita Y., Hicks J. C. Investigating the multifunctional nature of bimetallic FeMoP catalysts using dehydration and hydrogenolysis reactions. *Appl. Catal. A.* Aug 25 2016;524:85-93.
3. Rensel, D. J., Kim J., Jain V., Bonita Y., Rai N., Hicks J. C. Composition-directed $Fe_xMo_{2-x}P$ bimetallic catalysts for hydrodeoxygenation reactions. *Catal. Sci. Technol.* 2017;7:1857-1867.
4. Bonita, Y., Hicks J. C. Periodic trends from metal substitution in bimetallic Mo-based phosphides for hydrodeoxygenation and hydrogenation reactions. *J. Phys. Chem. C.* 2018, ASAP.

532119 Etherification of 5-Hydroxymethylfurfural Using Zeolite Catalysts

Abstract:

Biomass is a promising feedstock for the production of modern chemicals. In this regard, polymers represent an attractive class of potential biobased materials, and it may be possible to synthesize a variety of polymeric materials starting from 5-hydroxymethylfurfural (HMF). In this study, we show that H-BEA zeolite is highly active and selective for the production of a variety of HMF ethers which would be suitable for subsequent polymerization.

Several catalysts were evaluated for their etherification activity, including: H-BEA and H-ZSM-5 zeolites, Amberlyst-15 (A-15), and Î³-Al_2O_3. These catalysts were chosen to evaluate a variety of active sites and pore structures. Catalysts possessing BrÃ¸nsted acid sites (*i.e.*, H-ZSM-5 and H-BEA zeolites, and A-15) showed the highest selectivities to the ether product, with H-BEA achieving nearly quantitative selectivity. We hypothesize that the improved selectivity observed for H-BEA, compared with A-15, is due to confinement effects that inhibit both levulinate formation and biomolecular condensation (*i.e.*, resinification) reactions.

The rate of etherification is nearly constant for the series of ethanol, *n*-butanol, cyclohexanol, and phenol. Moreover, while the etherification rate saturates at high HMF concentrations, there is little variation in the rate across for alcohol concentrations ranging from 0.1 to 10 mol L^{-1}. These kinetics data, coupled with the invariance of the rate with varying alcohol data, suggest that the reaction proceeds through an oxocarbenium ion transition state that does not involve the etherifying alcohol.

532366 Mechanistic Insights into the Hydrogenolysis of Levoglucosanol over Bifunctional Platinum Silica-Alumina Catalysts in Tetrahydrofuran Solvent

Abstract:

The cellulose-derived platform molecule levoglucosenone (LGO) can be converted to a variety of high-value oxygenated chemicals. [1] LGO can be quantitatively hydrogenated to levoglucosanol (Lgol) over metal catalysts. [2] Herein, we report on the hydrogenolysis of Lgol to tetrahydrofurandimethanol (THFDM) and tetrahydropyran-2-methanol-5-hydroxyl (THP2M5H) over bifunctional platinum or palladium catalysts supported on silica-alumina in tetrahydrofuran solvent. [3,4] THFDM is an α,ω-diol and potential polymer precursor, while both THFDM and THP2M5H can be further upgraded to the commodity chemical 1,6-hexanediol.

A combination of experimental reaction kinetics studies and first principles calculations were used to investigate the mechanism and the roles of the metal and acid sites in this reaction. THFDM and THP2M5H are produced in parallel pathways, and the ketone intermediate THP2M5one is a precursor to THP2M5H. [3] The THFDM cis/trans ratio is independent of the Lgol threo/erythro ratio, suggesting that the mechanism passes through an acyclic intermediate which erases the stereochemistry of the Lgol reactant. ^{13}C radio-labeling was used to confirm the ring rearrangement forming THFDM. The catalyst undergoes moderate, reversible deactivation with time-on-stream in a continuous flow reactor. The reaction rates and product selectivities are comparable at 1.1% and 5.3% Pt loadings, indicating that the reaction rate is acid-limited at these metal loadings; the metal-catalyzed hydrogenation step is not rate-limiting (Figure 1). [4] The measured zero-order dependence in hydrogen indicates that the rate-limiting acid-catalyzed step precedes the fast hydrogenation step. As the Pt loading in Pt/SiAl catalysts is decreased, or when the bare SiAl support is separated from a Pt/SiO_2 catalyst in a dual-layer configuration, the selectivity towards identified products decreases (Figure 1). These results suggest that degradation reactions are avoided when the reactive intermediates formed over acid sites are rapidly hydrogenated over metal sites. First principles simulations were performed to investigate the energetics of the proposed reaction pathway consistent with the experimental results.

Figure 1. *Lgol consumption rate (left y-axis) and product selectivity (right y-axis) versus Pt loading for Pt/SiO_2-Al_2O_3 catalysts.*

These insights into the mechanism of Lgol hydrogenolysis provide a foundation for rationally designed processes to produce high-value chemicals from lignocellulosic biomass using bifunctional metal-acid catalysts.

References.

1. Cao, F.; Schwartz, T. J.; McClelland, D. J.; Krishna, S. H.; Dumesic, J. A.; Huber, G. W., Dehydration of cellulose to levoglucosenone using polar aprotic solvents. *Energy & Environmental Science* **2015,** *8* (6), 1808-1815.

2. He, J.; Huang, K.; Barnett, K. J.; Krishna, S. H.; Alonso, D. M.; Brentzel, Z. J.; Burt, S. P.; Walker, T.; Banholzer, W. F.; Maravelias, C. T.; Hermans, I.; Dumesic, J. A.; Huber, G. W., New catalytic strategies for [small alpha],[small omega]-diols production from lignocellulosic biomass. *Faraday Discussions* **2017,** *202* (0), 247-267.

3. Krishna, S. H.; McClelland, D. J.; Rashke, Q. A.; Dumesic, J. A.; Huber, G. W., Hydrogenation of levoglucosenone to renewable chemicals. *Green Chemistry* **2017,** *19* (5), 1278-1285.

4. Krishna, S. H.; Assary, R. S.; Rashke, Q. A.; Schmidt, Z. R.; Curtiss, L. A.; Dumesic, J. A.; Huber, G. W., Mechanistic Insights into the Hydrogenolysis of Levoglucosanol over Bifunctional Platinum Silica–Alumina Catalysts. *ACS Catalysis* **2018**, 3743-3753.

538548 Hydrothermal Stability of Chloromethyl Polystyrene Based Solid Acid Catalysts and Mechanism of Cellulose Hydrolysis

Abstract:

Cellulose is one of the major components of lignocellulosic biomass and the most abundant renewable polymer on earth. As such it could serve as a sustainable feedstock for the production of commercially important fuels and chemicals. However, cellulose conversion to smaller platform molecule is the major bottleneck for economically competitive utilization of lignocellulosic biomass.

The most common processes to depolymerize cellulose employ liquid acid or enzyme catalyzed hydrolysis to produce glucose. However, issues such as reactor corrosion, slow reaction rates, and inability to recycle limits the cost-effective conversion of cellulose. Alternatively, solid acids have been proposed as reusable catalysts for cellulose hydrolysis. Introduction of binding groups in the catalyst structure is hypothesized to increase the affinity between the solid acid and cellulose, brining the glycosidic bonds within contact with catalytic sites resulting in a higher hydrolysis rate. This hypothesis finds support in adsorption studies of small cellulosic molecules such as cellobiose and glucose. However, the mechanism of such catalytic activity has not been elucidated.

Chloromethyl polystyrene based bifunctional catalysts have showed high activity towards cellulose depolymerization. It is hypothesized the chlorine participates in hydrogen bonding with the hydroxyl groups in cellulose chains. However, the chloromethyl groups is rather label and could hydrolyze in hydrothermal conditions generating hydrochloric acid. Hydrochloric acid could also catalyze cellulose depolymerization, but its effect on the overall catalytic mechanism has not be evaluated.

In this work we studied partially sulfonated chloromethyl polystyrene catalyst for cellulose hydrolysis. Structural analysis of catalyst revealed that synthesis procedure results in little to no chlorine available on the external surface, but presence of hydroxyl groups, which can also participate in hydrogen bonding with cellulose. The catalyst showed high activity towards hydrolyzing cellulose resulting in levulinic acid and formic acid as major products. However, analysis of the solid and the liquid media revealed chloromethyl group hydrolysis and hydrochloric acid generation. This observation raised the question if the catalyst is indeed acting as a solid acid.

In order to elucidate the mechanism of hydrolysis we subjected the catalyst to hydrothermal reaction conditions in the absence of cellulose. Monitoring pH and chloride ion concentration, we developed a kinetic profile of hydrochloric acid generation. Cellulose hydrolysis kinetic models involve ion concentration as a preexponential factor on the expression for the reaction rate constant. We hypothesized that incorporating the time dependence of the acid concentration in the kinetic model of cellulose hydrolysis to glucose and subsequent dehydration to HMF, levulinic acid, and formic acid would be able to accurately predict experimental results. An acceptable agreement between model and experimental product yields pointed to the liquid acid acting as the primary catalyst, obscuring any catalytic activity, if any, of the solid catalyst. Further support was obtained by carrying out hydrolysis by a titrated solid acid catalyst, in which the sulfonic acid groups have been ion exchanged and, thus, rendered inactive; similar yields were obtained as the non-titrated catalyst. The results in this investigation highlight the need for systematic study of proposed solid acids for cellulose hydrolysis and consideration of potential release of homogeneous acids in the reaction media.

538054 Glycerol Transfer-Hydrogenation of Levulinic Acid Using Ru and Ir Carbene Organometallics Immobilized on Active Hydrotalcites

Abstract:

The mechanism and kinetics of levulinic acid (LA) hydrogenation to gamma-valerolactone (GVL) have been extensively studied over the last decade. Despite improvements in selectivity and the use of alternative hydrogen sources such as formic acid and secondary alcohols, economic production of GVL remains hindered by slow reaction rates and the high cost of current hydrogen sources. The use of glycerol, a waste byproduct in biodiesel synthesis, as a hydrogen donor could alleviate the economic challenges associated with GVL production. Herein we tested hydrogenation/deoxygenation of LA under dilute aqueous to neat conditions using Ru and Ir N-heterocyclic carbene (NHC) organometallic catalysts while also varying pH with potassium hydroxide. A total of 7 Ru and Ir homogeneous catalysts were screened for activity, the most active catalysts were then immobilized onto tunable hydrotalcites (HT). A 3:1 ratio of Mg:Al HT was used as a basis for all HT catalytic reactions while substitutions of Al with Fe, Ni, and Co were tested for improved catalytic activity. Products were identified using LCMS, HPLC, NMR, GCMS, and FTIR. The products formed and reaction rates were pH-dependent; acidic conditions favored ketal formation between LA and glycerol whereas alkaline conditions with an Ir(I) complex favored GVL and lactic acid formation at a turnover frequency (TOF) above 66,000 h^{-1} at 150ËšC for GVL. The Ir(I) catalyst achieved unprecedented reaction rates with glycerol, therefore, a comparison to reaction rates achieved with formic acid was demonstrated and showed negligible differences in rates and selectivity. Upon immobilizing the NHC catalysts on HT, the non-active Ru and Ir catalysts achieved activity over 10,000 h^{-1}. Under the same conditions the immobilized Ir(I)HT rate was reduced to 15,000 h^{-1} for GVL. The increase in activity from the other Ru and Ir(III) complexes tested are believed to be a result of the HT Lewis acid sites.

530511 Paired Electrocatalytic Hydrogenation and Oxidation of 5-Hydroxymethylfurfural for Efficient Production of Biorenewable Monomers

Abstract:

Electrochemical synthesis is a promising route for sustainable chemical production; however, its widespread application is hindered by economic limitations incurred from high capital and operating costs. In this presentation, we report highly efficient electrocatalytic conversion of 5-(hydroxymethyl)furfural (HMF) to biobased monomers by pairing HMF reduction and oxidation half reactions in a single electrochemical cell. HMF hydrogenation to 2,5-bis(hydroxymethyl)furan (BHMF) was achieved under mild electrolyte conditions and ambient temperature using a Ag/C cathode. Meanwhile, HMF oxidation to 2,5-furandicarboxylic acid (FDCA) with ~100% efficiency was facilitated under the same conditions by a homogeneous nitroxyl radical redox mediator, together with an inexpensive carbon felt anode. The selectivity and efficiency for Ag-catalyzed BHMF formation were sensitive to cathode potential, owing to competition from HMF hydrodimerization reactions and water reduction (hydrogen evolution). Moreover, the carbon support of Ag/C was active for HMF reduction and contributed to undesired dimer/oligomer formation at strongly cathodic potentials. As a result, high BHMF selectivity and efficiency were only achieved within a narrow potential range near â€“1.3 V. Fortunately, the selectivity of redox-mediated HMF oxidation was insensitive to anode potential, thus allowing HMF hydrogenation and oxidation half reactions to be performed together in a single cathode-potential-controlled cell. Electrocatalytic HMF conversion in a paired cell achieved high molar yields of BHMF and FDCA, and nearly doubled electron efficiency compared to the unpaired cells. Future research will be focused on rational design of an electrochemical flow reactor, in which the use of a membrane electrode assembly (MEA) can significantly lower the internal resistance and reduce the total energy demand.

525127 A Full FurfuralÂ Utilization over Ni/SiO_2 Catalysts

Abstract:

Furfural can be mass-produced from lignocellulose biomass and can be a platform chemical for producing valuable chemicals. In this study, we examine Ni/SiO_2 catalysts for the conversion of furfural under a hydrogen atmosphere. The reaction products include furfuryl alcohol (FA), furan, and propylene as major; butanal and other hydrogenated products, indicating the reaction routes involving side chain hydrogenation, decarbonylation, and ring opening. The reactivity and the product selectivity depend on the reaction temperature and the Ni particle size. The pretreatment by including calcination prior to hydrogen reduction leads to Ni/SiO_2â€“CR with large Ni particles (~ 15 nm) and a high selectivity to FA at below 200 °C. The Ni/SiO_2â€“R is hydrogen-pretreated without the calcination and it contains small Ni particles (~ 5 nm) and exhibits relatively high selectivity to furan. Moreover, the furan can be reacted seriously to produce propylene and CO by this Ni catalyst. The turnover frequency (TOF) of furfural conversion is 239 and 408 h^{-1}, respectively, on Ni/SiO_2-CR and Ni/SiO_2-R at 175 °C, when the former shows 100% selectivity to FA and the latter exhibits a selectivity of around 38% and 62%, to FA and furan, respectively. The results indicate that the furfural hydrotreating reaction over Ni catalysts is with structure sensitive and give a full furfural utilization, either to FA (a biofuel), or to propylene and CO (a chemical feedstock).

532256 Multi-Objective and Dynamic Real-Time Optimization of Postcombustion Carbon Capture Processes for Cycling Applications

Abstract:

As the penetration of intermittent renewable energy sources, such as solar and wind, into the power grid increases[1], there is a growing need for baseload fossil-fueled power plants to cycle their load accordingly[2]. Cycling baseload plants may introduce unexpected inefficiencies into the system and increase costs, as the power plants are no longer operating at their optimal design point. Previously[3], a dynamic real-time optimization (DRTO) framework was implemented on a reduced-order model of a methylethylamine (MEA)-based carbon capture subsystem (CCS) from a supercritical coal-fired power plant[4]. This framework considered the cycling profile, electricity price market and *cap&trade* as forcing functions to determine the optimal output trajectories for different carbon capture landscapes. The obtained optimal dynamic trajectories can then ultimately be sent to lower-level advanced model-based controllers[5].

Although the implementation of the DRTO framework for the CCS was performed successfully, in general, real-life situations are guided by more than a stand-alone economic objective to determine the optimum point at a certain time. Particularly, for the cycling scenario, objectives just as important such as emission reduction or minimization of wear and tear may be conflicting with the economic cost. To fill this gap, it is proposed in this work the incorporation of a multi-objective optimization (MOO) component into the DRTO framework for the calculation of optimal output trajectories considering all the objectives in tandem. In this case, more reliable optimal trajectories can be obtained, as the framework considers a more comprehensive context. As a trade-off, there is an increase in algorithm complexity, as new definitions (e.g., dominance) need to be introduced, and different strategies (e.g. aggregation, Pareto-front) need to be assessed to achieve the solution.

The reduced-order model of the subsystem considered in this work is a nonlinear autoregressive with exogenous inputs (NARX) with wavelets networks. The DRTO calculates the output optimal trajectories considering a one day-ahead prediction with a 1 hour simulated resolution. For the MOO component, it is desirable to minimize the decision maker interaction with the optimizer, so that the optimal trajectories can be determined and implemented at every time interval in an online and automated fashion. For instance, the optimizer should not wait for the decision maker to choose a point among a set of solutions to send the trajectories to the controllers. Thus, a multi-objective framework that guarantees a Pareto-optimal operating point is implemented in this work to keep the optimization independent of the hourly input of the decision maker. Furthermore, an algorithm without the necessity for system/search space convexity is also explored. The augmented framework is applied to a case study of the cycling carbon capture subsystem and the results are generated considering different carbon capture landscapes.

References

[1] REN21. (2018). Renewable Energy Policy Network for 21st Century. Available at: http://www.ren21.net/status-of-renewables/global-status-report/. Accessed on April 16, 2018.

[2] BENTEK ENERGY LLC. (2010). How Less Became More: Wind, Power and Unintended Consequences in the Colorado Energy Market. *Prepared for Independent Petroleum Association of Mountain States*.

[3] KIM, R., LIMA F. V., (2017). Nonlinear System Identification and Dynamic Real Time Optimization of Postcombustion CO_2 Capture Processes for Cycling Applications. *AIChE Annual Meeting, Minneapolis, MN.*

[4] ZHANG Q., TURTON R., & BHATTACHARYYA D., (2016). Development of Model and Model-Predictive Control of an MEA-Based Postcombustion CO_2 Capture Process. Industrial and Engineering Chemistry Research. 55, 1292-1308.

[5] HE X, WANG Y, BHATTACHARYYA D, LIMA F. V, & TURTON R., (2018). Dynamic Modeling and Advanced Control of Post-Combustion CO_2 Capture Plants. Chemical Engineering Research and Design. 131, 430-439.

536204 Dynamic Modeling and Control of a Post-Combustion Solid-Sorbent Capture System with the Ccsi Models and Tools

Abstract:

Solid sorbent-based post-combustion CO_2 capture processes are a viable alternative to solvent-based capture that may provide better energy efficiency. Integrated dynamic models of these types of processes are essential for understanding transient system behavior and for the development of control strategies to efficiently reject process disturbances and optimize tracking of set points. In this work, a suite of tools and models from the Carbon Capture Simulation Initiative (CCSI) was used to develop a dynamic integrated solid sorbent, post-combustion capture model and implement Nonlinear Model Predictive Control (NMPC). The CCSI Toolset has been released under an open-source license with the goal of accelerating and reducing scale-up risk of carbon capture technologies.

The post-combustion carbon capture flowsheet was developed in Aspen Custom Molder (ACM) and was constructed using the CCSI bubbling fluidized bed (BFB) model[1]. The model is one-dimensional with two-phases and an integrated heat-exchanger that is flexible over a wide-range of operating conditions so that it can represent both adsorption and regeneration in either a steady-state or dynamic framework. A dynamic flowsheet was developed that is based on an optimal steady-state design that includes a three stage BFB for CO_2 adsorption, a two stage BFB regenerator, and includes a crossflow heat exchanger between the rich and lean solid streams. The flowsheet also accounts for the pressure flow dynamics and downcomers in the BFB.

The integrated model is computationally expensive and not suitable for direct use in an NMPC framework, hence a reduced order model is required for model-based control. The CCSI data-driven system identification tool, D-RM Builder,[2] is used to generate computationally efficient dynamic reduced-order models from a set of transient data. D-RM Builder allows for the selection of input/output variables, the generation of a sequence of step-changes, and interfaces with rigorous modeling platforms, such as ACM, to automatically run and collect step-test data. This tool was used to generate a computationally efficient nonlinear Decoupled A-B Net (DABNet) model[3] involving multiple time-scales corresponding to "fast" pressure-flow dynamics and "slower" temperature-composition transients. The DABNet model was used as the control model in CCSI's Advanced Process Control (APC) Framework[4,5]. The APC Framework solves the constrained NMPC problem in a computationally efficient manner utilizing "fast" analytical derivative evaluations, implementing the control moves on the rigorous ACM flowsheet model. The control strategy is tested and compared to traditional control for its servo control and disturbance rejection characteristics.

References

1. Lee, A., & Miller, D. C. (2012). A one-dimensional (1-d) three-region model for a bubbling fluidized-bed adsorber. *Industrial & Engineering Chemistry Research*, *52*(1), 469-484.
2. Ma, J., Mahapatra, P., Zitney, S. E., Biegler, L. T., & Miller, D. C. (2016). D-RM Builder: A software tool for generating fast and accurate nonlinear dynamic reduced models from high-fidelity models. *Computers & Chemical Engineering*, *94*, 60-74.
3. Sentoni, G. B., Biegler, L. T., Guiver, J. B., & Zhao, H. (1998). State space nonlinear process modeling: Identification and universality. *AIChE Journal*, *44*(10), 2229-2239.
4. Omell, B. P., Ma, J., Mahapatra, P., Yu, M., Lee, A., Bhattacharyya, D., Zitney, S., Biegler, L., & Miller, D. C. (2016). Advanced Modeling and Control of a Solid Sorbent-Based CO2 Capture Process. *IFAC-PapersOnLine*, *49*(7), 633-638.
5. Mahapatra, P., Ma, J. & Zitney, S.E. (2018). Nonlinear Model Predictive Control using Decoupled A-B Net Formulation for Carbon Capture Systems – Comparisons with Algorithmic Differentiation Approach. *American Control Conference*, Jun 27-29, Milwaukee, WI.

537271 Application of a Data-Driven Modeling Approach to a Large-Scale Power Plant

Abstract:

Most conventional fossil-fuel-based power plants are designed to operate in baseload operation with small variations in the power output [1]. With the increasing contribution of renewable energy sources to the power grid, the conventional power plants are experiencing more fluctuations in the grid. This is mainly because most of the renewable energy sources are intermittent in nature [2] and conventional power plants must ramp up/down to provide room for the renewable power to be utilized or to compensate for them when they become unavailable. The sharp ramps in the power output of conventional sources result in increased inefficiencies in the power plant operation, which results in increased generation of pollutants such as NOx. It is, therefore, critical to optimize the operation of the power plant when it undergoes transient circumstances.

In this research study, a large scale coal-fired power generation unit in the United States was considered to demonstrate the benefits of dynamic optimization, as opposed to the steady-state optimization approach that is currently implemented in this unit. Currently, power production from renewable sources as well as the electricity demand are forecasted a few hours ahead of time. This serves as an opportunity to plan the operation of the power plant ahead of time such that NOx generation in the plant is minimized while it undergoes fluctuations. The first step in implementing a dynamic optimization approach is to develop a dynamic model of the plant. This presentation provides the results of a dynamic model of the plant as opposed to a steady-state approach that is the currently utilized in the plant. The power plant considered in this analysis has hundreds of variables that makes it very complex to develop a first-principles model of the plant. Thus, a data-driven modeling approach was considered in this study. The method used to develop the dynamic model of the plant is based on a recurrent neural network (RNN) [3]. An RNN model uses neurons that have nonlinear activation functions, which is used to map inputs to outputs. An RNN model takes advantage of some of the past values of measurements (e.g. NOx in this study) that are fed back to the model; thus, making it dynamic in nature. Measurement feedback also serves as a recalibration of the model. An RNN model is also capable of predicting into the future of the plant for as many time steps ahead as there is forecasted data for power demand and renewable power sources.

To develop an RNN model of the plant, it should first be trained i.e. the different weighting and bias factors should be calibrated such that the model predictions for target outputs (i.e. NOx in this analysis) should have the minimum deviations with the corresponding historical data. To achieve this goal, historical data for all decision variables and target NOx values for over four months of the plant operation was used. Additionally, different numbers of neurons as well as model orders were investigated to balance the model prediction accuracy and its complexity. The order of an RNN model refers to the number of past measurements that are fed-back to the model. To compare the accuracy of the models with different neurons and orders, coefficient of correlation (R) was used as the performance metric.

The optimal number of neurons that can best represent the nonlinear system of this power plant was obtained after a sensitivity analysis. Subsequently, the training of the RNN was implemented for orders 1 to 3. Additionally, a static neural network was also trained with the same historical data to provide a comparison basis for the RNN models. In a static neural network, none of the past values of NOx is given to the model; thus, essentially making the neural network a steady-state model. Training of all models was implemented in the MATLAB programming language. The R-values associated with the static and RNN models of order 1 to 3 were observed to be 0.85962, 0.92919, 0.9295, and 0.92816, respectively. While all models show a strong relationship between the model prediction and historical data, the R-value associated with a recurrent neural network was always greater than 90 and greater than a static neural network. This demonstrates that a dynamic model can better represent the historical data than a static model. Among the different orders considered for the recurrent neural network, a marginal improvement was also seen by increasing the model order. Because increasing the model order increases the computational time to train and use the neural network model, a model order of 2 was selected for the further analysis of this study to balance the accuracy and computational complexity of running an RNN model.

The trained neural network was then used to predict NOx for future time steps. A sensitivity analysis was implemented on the length of the prediction horizon for both the static and RNN models. Consequently, it was observed that smaller prediction horizons resulted in higher and lower R- and root mean square error (RMSE) values, respectively, when compared to the historical NOx data; i.e. the model can more accurately predict the future operation of the plant with shorter horizons. This conclusion was expected because the further the model has to foresee into the future, it has to predict additional fluctuations that the actual plant has undergone through this period. A 3-hour prediction horizon was therefore selected for this analysis, which resulted in accurate prediction of NOx over the horizon while also enabling to demonstrate the benefits of dynamic simulation in comparison to a static model. In this study, historical data for the gross power output of the plant was used as a given input to the model for the desired prediction horizon and the NOx prediction of the model was compared with the historical NOx generation of the plant during the same period. While the static model does not require any measurement update for the 3-hour prediction horizon, this analysis assumed that the actual NOx data from the plant was available to the RNN model every 5 time steps (with 5-minute interval). For a 3-hour ahead prediction of NOx, the R-value for the static model was obtained to be 0.7796 while a value of 0.8298 was observed for the second order RNN model. This difference between the static and second order RNN models represents a 6.44% improvement in the accuracy of model prediction in favor of the RNN model. Improvement of the predictions with a RNN model became more obvious when a 24-hour prediction horizon was considered. With a 24-hour prediction horizon, it was observed that the R-value of the RNN model improved 26.92% over the static model. This is because the plant experiences more transient circumstances over a 24-hour horizon and the RNN model better encounters for these transient operations over this period. These results demonstrate the value of dynamic simulation in comparison to steady-state simulation, when the plant undergoes transient operations such as those encountered when renewable sources are being utilized in the power grid. The future research in this study includes developing a dynamic optimization algorithm for the transient operation of the plant as well as improving the accuracy of the second order RNN model.

References

[1] "Dynamic optimization of a hybrid system of energy-storing cryogenic carbon capture and a baseline power generation unit," Applied Energy, pp. 66-79, 2016.

[2] "Intermittency and Value of Renewable Energy," Journal of Political Economy, pp. 124, (4), 2016.

[3] Principles of Artificial Neural Networks, World Scientific Publishing Company, 2013.

531756 Model-Based Analysis of a Thermofluidic Engine for Low-Grade Heat Recovery: Accounting for Irreversible Thermal Losses

Abstract:

The Non-Inertive-Feedback Thermofluidic Engine (NIFTE) is a two-phase thermofluidic oscillator that can convert low-grade heat into useful mechanical work. It consists of one hot and one cold heat exchangers alongside several compartments as illustrated on Fig. 1. The working fluid undergoes cyclic condensation and vaporization phases by contacting with the heat exchangers, which induce sustained fluid oscillations. Most previous mathematical models proposed to analyse the dynamic behaviour of NIFTE have considered linear descriptions [1-4]. These linear models provide useful insights about the operation of NIFTE, e.g. in terms of the minimum temperature difference required, the oscillation frequency and exergetic efficiencies; but despite their ability to give accurate quantitative predictions, they fail to explain the sustained robust periodic oscillations that have been observed experimentally on a NIFTE prototype. This is the so-called limit-cycle behaviour, which is essentially a nonlinear phenomenon. Markides et al. [5] first proposed a nonlinear inertive NTP model for NIFTE, and this model has been validated experimentally. Later, Wang et al. [6] conducted an optimization-based analysis using this model and they predicted the co-existence of multiple cyclic steady states for certain NIFTE configurations. **Fig. 1.** A schematic diagram of the NIFTE device.

However, none of the previous (linear or nonlinear) models account for the thermal losses in the device, which can result in significant overestimation of the efficiencies reported in experiments. Solanki et al. [7,8] refined several linear models models by incorporating a dissipative thermal loss parameter to account for (i) the loss of exergy arising from irreversible heat transfer, and (ii) alternating parasitic condensation and re-evaporation within the engine. This refinement enables much closer predictions of the exergetic efficiencies of NIFTE, while having a limited effect on the predicted oscillation frequencies.

The main development in this work is a description of thermal losses in the nonlinear inertive NTP model of NIFTE to improve its predictive capability – namely, the inertive model with thermal loss (NTP-TL). We build on the electric-circuit analogy previously developed for the NTP model, by adding extra resistor and capacitor elements to describe the overall irreversible thermal losses. It is found that the NTP-TL model provides much smaller exergetic efficiencies in agreement with experimental observations, and also improves the oscillation frequency predictions compared with its linear counterparts. Finally, we apply an optimization-based approach based on NTP-TL to detect and characterize the cyclic steady states corresponding to various engine configurations, and we also conduct a (global) sensitivity analysis to determine those (combinations of) parameters having the greatest effect in terms of the exergetic efficiency. Overall, the NTP-TL model enables a more realistic understanding of the performance of the NIFTE prototype, and provides useful insights on how to design better prototypes.

References:

[1] Smith, T.C.B: Thermally driven oscillations in dynamic applications. PhD thesis. Cambridge: University of Cambridge, UK. 2006

[2] Markides, C.N. and Smith, T.C.B.: A dynamic model for the efficiency optimization of an oscillatory low grade heat engine. Energy, 36, 6967-6980, 2011

[3] Solanki, R., Galindo, A., and Markides, C.N.: Dynamic modelling of a two-phase thermofluidic oscillator for efficient low grade heat utilization: Effect of fluid inertia. Applied Energy, 89(1), 156-163, 2012

[4] Solanki, R., Galindo, A., and Markides, C.N.: The role of heat exchangeon the behaviour of an oscillatory two-phase low-grade heat engine. Applied Thermal Engineering, 53, 177-187, 2013.

[5] Markides, C.N., Osuosslale, A., Solanki, R., and Stan, G.B.V.: Nonlinear heat transfer processes in a two-phase thermofluidic oscillator. Applied Energy, 104, 958-977, 2013

[6] Wang, Y., Markides, C.N., and Chachuat, B.: Optimization-based Investigations of a Thermofluidic Engine for Low-grade Heat Recovery. Manuscript accepted by ADCHEM 2018

[7] Solanki, R., Mathie, R., Galindo, A., and Markides, C.N.: Modelling of a two-phase thermofluidic oscillator for low-grade heat utilisation: Accounting for irreversible thermal losses. Applied Energy, 106, 337-354, 2013

[8] Markides, C.N., Solanki, R., Galido, A.: Working fluid selection for a two-phase thermofluidic oscillator: Effect of thermodynamic properties. Applied Energy, 124, 167-185, 2014

537091 Model Predictive Control Designs to Achieve Uniform Growth of Simultaneously Propagating Multiple Fractures in Hydraulic Fracturing

Abstract:

Natural gas and oil resources are found in shale rock formations with ultra-low permeability. In ultra-low permeability formations, horizontal wells are typically drilled and each horizontal well is stimulated with massive fractures for enhanced recovery of oil and gas [1]. In multi-stage hydraulic fracturing treatments, simultaneously propagating multiple fractures with close spacing often induce non-uniform fracture development, resulting in one or two dominant fractures due to the uneven distribution of fracturing fluids [2]. One of the important contributing factors for the uneven distribution is the well-known phenomenon called ``stress shadow effects" [3]. In general, fracturing fluids are distributed to multiple fractures inversely proportional to their flow resistances, which is a function of perforation friction, wellbore friction and fracture propagation. The problem with the stress shadow effects is that it exerts extra compressional stresses on the interior fractures and increases the flow resistance within interior fractures, resulting in fluid diverting into the exterior fractures. Therefore, to achieve uniform growth of simultaneously propagating multiple fractures and facilitate more fracturing fluids entering into the interior fractures, the stress shadow effects should be balanced or mitigated.

In this regard, several hydraulic fracture simulators have been developed to describe the stress shadow effects in simultaneously propagating multiple fractures [4]. Typically, these high-fidelity hydraulic fracture simulators require several days, and sometimes over one week, to compute the growth of simultaneously propagating multiple fractures at real reservoir length and time scales. Hence, optimization of hydraulic fracturing to find the operating condition for uniform growth of hydraulic fractures, which may often require hundreds or thousands of simulation runs, were not practically viable; in these directions, very limited efforts have been made by selecting a few important parameters via sensitivity analysis [4, 5].

Motivated by our previous efforts on single fracture propagation [6, 7, 8], in this work, we focus on the development of a new model order-reduction technique for simultaneously propagating multiple fractures by integrating the analytical models to calculate the pressure drop due to perforation friction and wellbore friction and a data-based ROM to describe the pressure drop along the fractures due to stress shadow effects and fracture interaction. Then, we propose a model-based pumping schedule design technique by utilizing the ROM and the uniform limited entry design technique to compute the flow rate of fracturing fluids which will promote equal distribution of fracturing fluids to achieve uniform growth of multiple fractures while mitigating the undesired stress shadow effects. Simulation results are presented to compare the proposed technique with state-of-the-art techniques in the field.

References:

[1] Lecampion, B., & Desroches, J. (2015). Simultaneous initiation and growth of multiple radial hydraulic fractures from a horizontal wellbore. *Journal of the Mechanics and Physics of Solids*, 82, 235-258.

[2] Wu, K., & Olson, J. E. (2016). Mechanisms of simultaneous hydraulic-fracture propagation from multiple perforation clusters in horizontal wells. *SPE Journal*, 21(03), 1000-1008.

[3] Germanovich, L. N., Astakhov, D. K., Mayerhofer, M. J., Shlyapobersky, J., & Ring, L. M. (1997). Hydraulic fracture with multiple segments I. Observations and model formulation. *International Journal of Rock Mechanics and Mining Sciences*, 34(3-4), 97.e1-97.e19.

[4] Kan, W., Anusarn, S., & Jizhou, T. (2016). Numerical study of flow rate distribution for simultaneous multiple fracture propagation in horizontal wells. *In 50th US Rock Mechanics/Geomechanics Symposium. American Rock Mechanics Association (ARMA-2016-038),* Houston TX.

[5] Cheng, C., Bunger, A.P., Peirce, A.P., 2016. Optimal perforation location and limited entry design for promoting simultaneous growth of multiple hydraulic fractures. *In SPE Hydraulic Fracturing Technology Conference. Society of Petroleum Engineers (SPE 179158),* The Woodlands TX.

[6] Siddhamshetty, P., Yang, S., & Kwon, J. S. I. (2017). Modeling of hydraulic fracturing and designing of online pumping schedules to achieve uniform proppant concentration in conventional oil reservoirs. *Computers & Chemical Engineering. URL: https://doi.org/10.1016/j.compchemeng.2017.10.032.*

[7] Siddhamshetty, P., Kwon, J. S. I., Liu, S., & Valkó, P. P. (2017). Feedback control of proppant bank heights during hydraulic fracturing for enhanced productivity in shale formations. *AIChE Journal*, 64(05), 1638-1650.

[8] Narasingam, A., & Kwon, J. S. I. (2017). Development of local dynamic mode decomposition with control: Application to model predictive control of hydraulic fracturing. *Computers & Chemical Engineering,* 106, 501-511.

526564 Real-Time Control and Balancing of a Reformer Furnace

Abstract:

In a vision of carbon-emission-free and sustainable energy systems, the hydrogen economy, in which hydrogen is the major energy carrier of the renewable energy sources and is the main transportation fuel, is a viable option and has the potential to reduce greenhouse gas emissions and to resolve issues associated with climate change without halting economic growth. Therefore, solving challenges encountered in commercial-scale production of hydrogen and optimizing these production lines has become an issue of great interest to both academia and industry. Among a variety of commercial hydrogen production processes, steam methane reforming (SMR) is by far the most common. For instance, SMR was responsible for 80% to 85% of the world-wide hydrogen production in 2007 [1]. SMR is typically carried out in top-fired steam methane reforming furnaces which are the most expensive equipment in terms of the maintenance and operating costs compared to other major equipment such as the hydrotreating, prereforming, water-shift and purification units at centralized SMR-based facilities. For instance, the re-tubing cost of a furnace is 10% of the total capital investment, and the annual operating cost to procure fresh natural gas for a SMR-based hydrogen plant with a production rate of 2.7 million Nm^3 per day is $62 million [2]. In addition, the conversion of methane via SMR in the furnace can only be increased by increasing the total energy input, which is constrained by the extreme sensitivity of the reforming tube service life to its operating temperature [3]; consequently, the plant throughput is compromised to avoid premature failure of the reformer and to reduce the plant risk of suffering substantial capital and production losses. Hence, the need for a robust tool that identifies the optimized total fuel flow rate and the corresponding valve distribution such that the conversion of methane via SMR is maximized, and the maximum OTWT along the reforming tube length among all reforming tubes is strictly less than the design temperature of the reforming tube wall material becomes apparent.

Motivated by the above considerations, the present work utilizes the framework for the furnace-balancing scheme [3], the valve-to-flow-rate converter [3], the statistical-based model identification [4] and a heuristic search algorithm to create a real-time balancing procedure, which recursively tests different total fuel flow rates of which the respective spatial distribution to burners is optimized in real-time until key operational specifications are satisfied. The ability to adjust the total fuel flow rate of the balancing procedure is of special interest for the hydrogen manufacturing industry as it can potentially lead to substantial savings in the re-tubing cost of the reformer. Subsequently, a computational fluid dynamic (CFD) model of the furnace developed in [5] is used to characterize the previously unstudied dynamic behavior of the reformer, based on which we develop an optimal strategy to implement the optimized total fuel flow rate to maximize the reformer throughput. Finally, a case study in which the balancing procedure is implemented on the reformer initially operated under the nominal reformer input is proposed, and the results are used to demonstrate that the furnace-balancing scheme successfully determines the optimized reformer fuel input to increase the reformer throughput while meeting the OTWT limits.

[1] Simpson, A.P., Lutz, A.E., 2007. Exergy analysis of hydrogen production via steam methane reforming.

International Journal of Hydrogen Energy 32, 4811–4820.

[2] Latham, D.A., McAuley, K.B., Peppley, B.A., Raybold, T.M., 2011. Mathematical modeling of an industrial steam-methane reformer for on-line deployment. Fuel Processing Technology 92, 1574–1586.

[3] Tran, A., Aguirre, A., Crose, M., Durand, H., Christofides, P.D., 2017a. Temperature balancing in steam methane reforming furnace via an integrated CFD/databased optimization approach. Computers & Chemical Engineering 104, 185–200.

[4] Tran, A., Pont, M., Aguirre, A., Durand, H., Crose, M., Christofides, P.D., in press, 2018. Bayesian model averaging for estimating the spatial temperature distribution in a steam methane reforming furnace. Chemical Engineering Researchand Design.

[5] Tran, A., Aguirre, A., Durand, H., Crose, M., Christofides, P.D., 2017b. CFD modeling of a industrial-scale steam methane reforming furnace. Chemical Engineering Science 171, 576–598.

521550 Finite Element Modeling and Optimization of Heat Exchangers

Abstract:

The rise in energy demand throughout the globe calls for more energy efficient design using process intensification techniques. These designs require solving complicated steady state models describing the physical phenomenon happening inside the system. One of the most important and well-researched aspects of chemical plant design are heat exchangers, yet these are currently designed under the assumption that the physical properties of the streams remain constant inside the heat exchanger. This assumption breaks down significantly under phase change as liquid and vapor phase properties vary drastically and non-smoothly. Lack of *a priori* knowledge of where the phase transition occurs inside the heat exchanger posts a serious problem for the solution algorithms. This complexity is compensated by adding safety factors that lead to inefficient and over-design of heat exchangers thus decreasing the energy efficiency of the whole process.

In this work, a new optimization framework for heat exchanger design is presented that can solve the difficulty of non-smooth nature resulting from phase change inside the heat exchanger. With this in mind, we describe a discretized finite element structure [1], which can be used as a candidate heat exchanger model for large-scale process design problems including phase change. We model the non-smooth nature of phase-change using complementarity constraints [3] that are relaxed and solved iteratively until an acceptable tolerance level is achieved. The discontinuity in heat transfer coefficient arising from the huge difference in latent and sensible heat transfer is described by a smooth approximation. We apply the proposed model by designing heat exchangers in a single-stage and multi-stage Vapor Compression Refrigeration (VCR) cycle with R-152a (1,1-Difluoroethane) as the refrigerant. The VCR cycle subject to complementarity and process constraints with around 4000 variables and constraints is modeled as a non-linear program (NLP) in AMPL and solved using IPOPT [2]. The intricacies of the model along with initialization and the results from the optimization will be presented. Extensions to modeling the system dynamics and solving it using optimization solvers will be discussed.

[1] Ravikumaur, S.G., Seetharamu, K.N. and Aswatha Narayana, P.A., Finite element analysis of shell and tube heat exchanger, International Communications in Heat and Mass Transfer, Volume 15, Issue 2, Pages 151-163,(1986)

[2] Wächter Andreas, Biegler L.T., On the implementation of an interior-point filter line-search algorithm for large-scale nonlinear programming, Mathematical Programming, Volume 106, Issue1, Pages 25-57,(2006)

[3] B.T. Baumrucker, L.T. Biegler, MPEC strategies for optimization of a class of hybrid dynamic systems, Journal of Process Control, Volume 19, Issue 8, Pages 1248-1256(2009)

538305 Application of Paroc in the Optimization and Control of PEM Water Electrolysis Process

Abstract:

Proton Exchange Membrane Water Electrolyzer (PEMWE) has gained attention within the last decade due to its relative advantage over other water electrolysis technologies. However, its high operational energy intensity (losses) has been the major setback (1). Energy losses are categorized as faradaic losses, and overpotentials and they are caused by complex irreversible interactions within the PEMWE process. The bubble effect, which involves the accumulation of oxygen gas bubble on the surface of the Membrane Electrode Assembly (MEA), is one of the various phenomena that has a significant impact on the overpotential of the PEMWE process (2–4). Furthermore, the flow of the electrolyte has a substantial effect on the accumulation of bubbles on the surface of the MEA(3). Few researchers have considered the bubble effect in their models (5) and none has attempted to systematically optimize their model to obtain an optimal operating point.

In this work, we present an experimentally validated dynamic mathematical model and control optimization of a PEMWE system using the Parametric Optimization and Control (PAROC) framework (6). The main objectives of this work are to (i) formulate and validate a comprehensive model; (ii) to obtain optimal operating points that minimizes the energy losses in the PEMWE process; (iii) and to design advanced controller to maintain the optimal operating points obtained. The PAROC framework enables the representation and solution of demanding model-based operational optimization and control problems. The application of PAROC on PEMWE features: (i) a high-fidelity dynamic mathematical model of the PEMWE that captures the detailed electrochemical interaction, transport phenomenon, and other interaction associated with energy losses in the system; (ii) a model reduction step that develops a more tractable state-space model from the original complex model without losing its sense of the original model; (iii) a Model Predictive Control (MPC) design and its reformulation to a multi-parametric (mp) optimization problem (which becomes mp-MPC) (7). The solution of the mp-MPC provides a map of solution that spans the feasible operating region of the PEMWE. Unlike the MPC the mp-MPC avoids the online optimization procedure at every time step (7). The optimization is done once and offline to determine the control action at every realization of the parameters while simultaneously accounting for physical and operational constraints. For a system with fast dynamics (like the PEMWE), mp-MPC is ideal.

Reference

1. Grigoriev SA, Porembsky VI, Fateev VN. Pure hydrogen production by PEM electrolysis for hydrogen energy. Int J Hydrogen Energy [Internet]. 2006 Feb 1 [cited 2018 Jan 17];31(2):171–5. Available from: http://www.sciencedirect.com/science/article/pii/S036031990500145X
2. Qian K, Chen ZD, Chen JJJ. Bubble coverage and bubble resistance using cells with horizontal electrode. J Appl Electrochem [Internet]. 1998;28(10):1141–5. Available from: http://link.springer.com/10.1023/A:1003447913204
3. Balzer RJ, Vogt H. Effect of Electrolyte Flow on the Bubble Coverage of Vertical Gas-Evolving Electrodes. J Electrochem Soc [Internet]. 2003;150(1):E11. Available from: http://jes.ecsdl.org/cgi/doi/10.1149/1.1524185
4. Kiuchi D, Matsushima H, Fukunaka Y, Kuribayashi K. Ohmic Resistance Measurement of Bubble Froth Layer in Water Electrolysis under Microgravity. J Electrochem Soc [Internet]. 2006;153(8):E138. Available from: http://jes.ecsdl.org/cgi/doi/10.1149/1.2207008
5. Grigoriev SA, Kalinnikov AA, Millet P, Porembsky VI, Fateev VN. Mathematical modeling of high-pressure PEM water electrolysis. J Appl Electrochem. 2010;40(5):921–32.
6. Pistikopoulos EN, Diangelakis NA, Oberdieck R, Papathanasiou MM, Nascu I, Sun M. PAROC An integrated framework and software platform for the optimisation and advanced model-based control of process systems. Chem Eng Sci. 2015;136:115–38.
7. Bemporad A, Morari M, Dua V, Pistikopoulos EN. The explicit linear quadratic regulator for constrained systems. Automatica. 2002;38(1):3–20.

526054 Predictive Control of Solar Collector Energy System with Gaussian Process Priors of Uncertain Solar Irradiance

Abstract:

In this work, we study the application of the receding horizon optimal control for solar collector energy system governed by coupled nonlinear hyperbolic partial differential equations [1]. The optimal solution is calculated by using the calculus of variations approach. Solar irradiance is the only dominant source to provide heat to the system and suppose that current time solar irradiance can be measured by a solar irradiance measuring device such as pyranometer. In order to apply nonlinear model predictive control (MPC) techniques [2], the prediction of solar irradiance in future certain time horizon is necessary. In this work, we consider the multi-step ahead prediction of solar irradiance (time series) using non-parametric Gaussian process (GP) regression parametrized by mean and covariance function [3]. k-step ahead forecasting of a discrete-time non-linear dynamic system can be performed by doing repeated one-step ahead predictions. k-step ahead prediction method works as follows: it predicts only one time step ahead, using the estimate of the output of the current prediction, as well as previous outputs (up to the lag L), as the input to the prediction of the next time step, until the prediction k steps ahead is made. In the simulation part, the integration of GP model of solar irradiance and MPC is performed to achieve optimal outlet temperature trajectory tracking.

[1] Xu, Xiaodong, Yuan Yuan, and Stevan Dubljevic. "Receding horizon optimal operation and control of a solarâ €thermal district heating system." AIChE Journal (2017)

[2] Van Pham, Thang, Didier Georges, and Gildas Besancon. "Predictive control with guaranteed stability for water hammer equations." IEEE transactions on automatic control 59.2 (2014): 465-470.

[3] Rasmussen, Carl Edward. "Gaussian processes in machine learning." Advanced lectures on machine learning. Springer, Berlin, Heidelberg, 2004. 63-71.

AUTHOR INDEX

AUTHOR INDEX

AUTHOR INDEX

AUTHOR INDEX

AUTHOR INDEX